VARIABLE STARS IN GLOBULAR CLUSTERS
AND IN RELATED SYSTEMS

ASTROPHYSICS AND SPACE SCIENCE LIBRARY

A SERIES OF BOOKS ON THE RECENT DEVELOPMENTS
OF SPACE SCIENCE AND OF GENERAL GEOPHYSICS AND ASTROPHYSICS
PUBLISHED IN CONNECTION WITH THE JOURNAL
SPACE SCIENCE REVIEWS

Editorial Board

J. E. BLAMONT, *Laboratoire d'Aeronomie, Verrières, France*

R. L. F. BOYD, *University College, London, England*

L. GOLDBERG, *Harvard College Observatory, Cambridge, Mass., U.S.A.*

C. DE JAGER, *University of Utrecht, Holland*

Z. KOPAL, *University of Manchester, Manchester, England*

G. H. LUDWIG, *NASA, Goddard Space Flight Center, Greenbelt, Md., U.S.A.*

R. LÜST, *Institut für Extraterrestrische Physik, Garsching-München, Germany*

B. M. MCCORMAC, *Lockheed Palo Alto Research Laboratory, Palo Alto, Calif., U.S.A.*

H. H. NEWELL, *NASA, Washington, D.C., U.S.A.*

L. I. SEDOV, *Academy of Sciences of the U.S.S.R., Moscow, U.S.S.R.*

Z. ŠVESKA, *Freiburg in Breisgau, F.R.G.*

Secretary of the Editorial Board

W. DE GRAAFF, *Sterrewacht 'Sonnenborgh', University of Utrecht, Utrecht, Holland*

VOLUME 36

VARIABLE STARS
IN GLOBULAR CLUSTERS
AND IN RELATED SYSTEMS

PROCEEDINGS OF THE IAU COLLOQUIUM No. 21 HELD AT
THE UNIVERSITY OF TORONTO, TORONTO, CANADA
AUGUST 29–31, 1972

Edited by

J. D. FERNIE

David Dunlap Observatory, University of Toronto, Toronto, Canada

D. REIDEL PUBLISHING COMPANY

DORDRECHT-HOLLAND / BOSTON-U.S.A.

Library of Congress Catalog Card Number 73–83560

ISBN 90 277 0341 8

Published by D. Reidel Publishing Company,
P.O. Box 17, Dordrecht, Holland

Sold and distributed in the U.S.A., Canada, and Mexico
by D. Reidel Publishing Company, Inc.
306 Dartmouth Street, Boston,
Mass. 02116, U.S.A.

All Rights Reserved
Copyright © 1973 by D. Reidel Publishing Company, Dordrecht, Holland
No part of this book may be reproduced in any form, by print, photoprint, microfilm,
or any other means, without written permission from the publisher

Printed in The Netherlands by D. Reidel, Dordrecht

TABLE OF CONTENTS

FOREWORD VII
EDITOR'S NOTE IX

PART I / GENERAL PROBLEMS OF VARIABLES IN POPULATION II SYSTEMS

H. S. HOGG / Variables in Globular Clusters, the Common and the Rare (Review Paper) 3
B. V. KUKARKIN (read by M. W. Feast) / Some Problems in the Study of Variable Stars in Globular Clusters 8
L. ROSINO AND G. PINTO / Observations of Faint Globular Clusters 21
S. VAN DEN BERGH / Planetary Nebulae, Globular Clusters and the Galactic Halo 26
R. J. DICKENS and A. L. T. POWELL / An Abundance Analysis of Fehrenbach's Star (HD 116745) in Omega Centauri 29
L. DETRE and B. SZEIDL (read by J. Graham) / On the Nature of the 41-Day Cycle of RR Lyrae 31
S. VAN AGT / Variable Stars in Dwarf Spheroidal Galaxies (Review Paper) 35

PART II / RR LYRAE VARIABLES IN POPULATION II SYSTEMS

L. ROSINO / Observational Aspects of RR Lyrae Variables in Globular Clusters (Review Paper) 51
A. TERZAN and B. RUTILY / New Variable Stars in the Globular Cluster NGC 6401 68
A. TERZAN, B. RUTILY, and CH. OUNNAS / Variable Stars in NGC 4590 76
R. J. DICKENS and ROSALIND FLINN / Two-Colour Photometry of RR Lyrae Variables in NGC 6981 82
D. H. P. JONES / Intermediate Band Photometry of RR Lyrae Variables in ω Cen and 47 Tuc 84
E. H. GEYER (read by J. D. Fernie) / UBV Magnitudes and Colours of 62 RR Lyrae Stars in ω Centauri (NGC 5139) 88
A. WEHLAU and NICHOLAS POTTS / Two-Color Observations of RR Lyrae Stars in M14 95
V. P. GORANSKIJ, B. V. KUKARKIN, and N. N. SAMUS' (read by R. J. Dickens) / On the Changes of Periods of RR Lyrae Variables in the Globular Clusters M5, M53 and NGC 5053 101

A. J. WESSELINK / Period and Period Changes of RR Lyrae Variables in M15   104
E. P. BELSERENE / Period Changes as Evidence for Evolution in ω Cen   105
M. V. NORRIS / Short Period Variables in the Magellanic Cloud Cluster NGC 1466   113
J. A. GRAHAM / The RR Lyrae Stars in the Magellanic Clouds   120

PART III / SLOW VARIABLES IN POPULATION II SYSTEMS

M. W. FEAST / Observational Aspects of Slow Variables in Globular Clusters (Review Paper)   131
C. M. COUTTS / A Preliminary Investigation of Period Changes for W Virginis Stars in Globular Clusters   145
S. DEMERS / Mean Magnitudes and Colours of Six Cepheids in Three Red Globular Clusters of the Large Magellanic Cloud   150
T. LLOYD EVANS and J. W. MENZIES / Red Variable Stars in Metal Rich Globular Clusters   151
R. F. WING / Narrow-Band Photometry of Red Variables in Globular Clusters   164
J. MENZIES / The Variable Stars in NGC 6723   178
B. V. KUKARKIN and A. S. RASTORGOUEV (read by L. Rosino) / The Period-Luminosity Relation for Cepheids in Globular Clusters   180

PART IV / THEORETICAL CONSIDERATIONS OF POPULATION II VARIABLES

P. DEMARQUE / Variable Stars and Evolution in Globular Clusters (Review Paper)   187
T. S. VAN ALBADA and N. BAKER / On the Two Oosterhoff Groups of Globular Clusters   196
V. CASTELLANI, P. GIANNONE, and A. RENZINI / On the Interpretation of RR Lyrae Properties in Globular Clusters and in Other Population II Systems   197
V. P. GORANSKIJ (read by S. van den Bergh) / Possible Evolutionary Interpretation of the Dependences on the Diagrams for RR Lyrae Variables   207
J. G. MENGEL / The Evolutionary Status of Population II Cepheids   214
A. V. SWEIGART and P. DEMARQUE / Semiconvection and the RR Lyrae Variables   221
A. V. MIRONOV (read by L. Woltjer) / The Helium Abundance in the Envelopes of the Bluest RR Lyrae Stars in Globular Clusters and Dependence of Globular Cluster Variable Star Properties on Chemical Composition   229

FOREWORD

This volume contains the papers and discussions at IAU Colloquium No. 21 on Variable Stars in Globular Clusters and in Related Systems held in Toronto on the 29th, 30th and 31st August 1972. It was the intention of the organizers that this meeting should honour the life long work in this field of Professor Helen Sawyer Hogg. She has been continuously active in observational research on variables in globular clusters for 46 years and her catalogues and bibliographies as well as her research papers, review articles and IAU reports as chairman of the committee on variable stars in clusters are of fundamental importance to all workers in this field.

The scope of the colloquium covered both observational and theoretical aspects of the problem, including the relationship of variables to non-variable cluster members, the position of the variables in the HR diagram and their importance for problems of stellar evolution, empirical data on the variables, periods and period changes, and the relevant parts of pulsation theory. The meeting was particularly successful in bringing together observers and theorists. It will have achieved its object if it has shown both observers and theorists which are the problems most suitable for attack at the present time. The meeting clearly demonstrated the great importance of research on variables in globular clusters and related systems for our understanding both of stellar evolution and stellar pulsation.

The colloquium was organized by a scientific organizing committee consisting of: L. Rosino, B. V. Kukarkin, H. S. Hogg, L. Detre, N. Baker, S. van Agt and M. W. Feast (chairman) and a local organizing committee consisting of: S. P. S. Anand, D. A. MacRae and J. D. Fernie (chairman). Inevitably most of the work fell on this local organizing committee, especially on Professor Fernie, the chairman. We are most grateful to them for all their work behind the scenes. Professor Fernie has in addition undertaken the task of editing this volume. The meetings were held in the University of Toronto and we are much indebted to the authorities of the University, the astronomy department and the David Dunlap Observatory for inviting us to hold the meetings there and for making so many fine facilities available. Financial support from the National Research Council of Canada, the National Science Foundation (U.S.A.) and the International Astronomical Union are gratefully acknowledged.

M. W. FEAST

Radcliffe Observatory, Pretoria
September, 1972

EDITOR'S NOTE

There were two papers given at the Colloquium which are not included in this book: a review paper by N. Baker entitled *Pulsation Theory Relevant to Variable Stars in Globular Clusters*, and another by K. von Sangbusch and N. Baker entitled *Stability of Non-Linear Periodic Pulsations of RR Lyrae Stars*. Regrettably, circumstances prevented Norman Baker from preparing written versions in time for inclusion in this volume.

I should like to take this opportunity to thank the following persons: Jennie Fabian for her extensive retyping of manuscripts, Gretchen Hagen for organizing and collating the discussion records, and Sheila Smolkin for assisting in the checking of references.

J. D. FERNIE

Toronto, September 1972

PART I

GENERAL PROBLEMS OF VARIABLES
IN POPULATION II SYSTEMS

VARIABLES IN GLOBULAR CLUSTERS, THE COMMON AND THE RARE

HELEN SAWYER HOGG

David Dunlap Observatory, University of Toronto, Richmond Hill, Ontario, Canada

The title of this talk is really just a different phrasing from one I have used at several IAU meetings on the subject of numbers and kinds of variables in globular clusters. To furnish this material, I have finished the *Third Catalogue of Variables in Globular Clusters*. Since many of you are coming to this Colloquium with new information, the Catalogue is in draft form with a request that corrections and additions be given me by October 2, after which the draft will go to the printer.

The *First Catalogue of Variables in Globular Clusters* was published at this observatory in 1939 and the Second Catalogue in 1955. In 1966 appeared the excellent *Catalogue of Variables South of Declination* $-29°$ by Fourcade, Laborde and Albarracin, with splendid large prints of identification charts.

The globular clusters searched for variables now total 105, and the number of variables found has passed the two thousand mark, to 2057. These numbers compare with 72 clusters searched and 1421 variables found, listed in the Second Catalogue. In 1955 I was surprised when the most frequent number of variables per cluster examined turned out to be one! This summer I had another surprise from the new tabulations. In the clusters searched to date, the most frequent number of variables per cluster is – zero! Twelve of the clusters searched have no variables. And the three variables in NGC 6397 are all described as field stars, so if these are excluded that would make a 13th cluster with no variables. Of the clusters with no variables, 6 are the deep southern clusters recently searched by Fourcade and Laborde. Time will tell how much selection effect is involved here. Naturally the clusters searched now will mostly be the more distant or highly obscured.

There are now five clusters each of which has more than 100 variables. These are Messier 3 with 212, Omega Centauri with 175, IC 4499 with 129, Messier 15 with 111 and Messier 5 with 102. As the survey becomes more complete, those with large numbers of variable stars stand out as rather rare systems. It is fortunate that the rich ones were among the earliest investigated, now 80 years ago by Prof. S. I. Bailey of Harvard, otherwise the project of hunting for variables in globulars might not have had so much impetus.

Of course RR Lyrae stars greatly predominate among the variables discovered. The statement I made years ago that roughly 90 percent of the variables in globular clusters are RR Lyrae stars still seems to hold. Periods have now been determined for 1157 RR Lyrae type in 46 clusters, providing much material for statistical correlations. The distribution of frequency of periods is not markedly different from that in 1955. There is a strong preference for periods between 0.28 and 0.40 day, and 0.46 through 0.66 day. Nearly 95 percent of the RR Lyrae periods in globular clusters fall within those intervals, with 23 percent in the shorter and 71 percent in the longer interval. Further significance is obtained when the frequency is studied cluster by cluster and correlated with various known properties of the cluster, as for example the correlation of Oosterhoff's Groups I and II with metallicity. Ever since Dr Martin Schwarzschild put the cluster variables in their place, that is in their gap in Messier 3, the processes which involve RR Lyrae stars have become better understood all the time.

At the long end of the RR Lyrae periods, an interesting anomaly is the number with longer periods in Omega Centauri. That cluster has 13 with periods longer than 0.75 days, more than twice as many as in all other globular clusters combined. And at the short end, two dwarf Cepheids have been detected, but probably neither is a cluster member. One in Omega Centauri has a period of 0.06 day and one in Messier 56 has 0.07 day.

Period changes are an important aspect of these stars, and are indicated in the catalogue merely by a + or − sign. Values of Beta for the same star by different investigators sometimes differ widely. Furthermore, Dr B. V. Kukarkin considers Beta to be a rather illusory quantity, and thinks that the best way of understanding the behaviour of the star is by O − C diagrams for different epochs. Period changes as well as the important aspect of colours will be discussed in later papers at this colloquium.

There are now 169 variables indicated not to be RR Lyrae stars in 44 clusters. I have arbitrarily divided them by period length. In the 1–30 day category, total 29, are the W Virginis stars and a few Cepheids with periods under a week. The 31–99 day category, total 23, includes a few RV Tauri stars, but is mainly made up of semiregular or cyclical variables. If no numerical value of a cycle is given, the star is assigned to the less definite irregular or semiregular category, which totals 27. In the range from 100 to 220 days, total 23, is found the type of long period variables which Dr Feast and his colleagues have been so diligent in proving to be actual cluster members. By contrast none of the 12 variables with periods over 220 days has been shown to be a cluster member and most of them have been proven field stars. Feast's limit of 220 days for the period of a globular cluster member still holds, but will it last through this colloquium? Clusters that have variables usually have both the RR Lyrae and slow types. It is rare for a cluster to have only one or the other. The cluster which really stands out with its richness in non-RR Lyrae variables is Omega Centauri. It has 17 in all, but not all of them are members. There are 7, all members, in the 1 to 30 day category, 5 in the 31 to 99 day, some of which are field, one 149 day, one field star at 235 days and 3 eclipsing stars, at least one of which is a member. Variable No. 1, the brightest variable in the cluster which has been considered for years to be RV Tauri type is now assigned a 29-day period with intermediate RV Tauri characteristics, by D. H. P. Jones. Great credit is due to the Herstmonceux observers for their massive pieces of valuable work on this cluster.

The recent detection of more irregular or semiregular variables in clusters constitutes a very important addition to our understanding of the stellar content of these. To a great extent this has come about through the marked increase in colour magnitude diagrams and colours available for individual stars. No one, I think, is going to try to fit an RR Lyrae period to a star near the tip of the red giant branch in a cluster, but in the past, when we did not know the star belonged there, we might have spent hours trying to do it. The multi-color work of O. J. Eggen has shown that many of the irregular variables in clusters lie in just such a location, and he goes so far as to suspect that all stars in clusters with $B-V$ greater than $+1.6$ will prove to be red irregular variables.

Some stars announced as variable by early workers have since been shown to be non-variable. But in recent years there has been a surprising twist to this, namely that three such stars have been shown to be really variable! These are in Messier 3, Nos. 8 and 15 by Kholopov, and No. 138 by Russev.

There are only about a dozen-and-a-half variables in globular clusters which do not fall into the foregoing categories of pulsating stars. Eclipsing systems represent about two thirds of these. One appears to be a member of NGC 3201 and at least one (No. 78) out of the three in Omega Centauri is a member. Sistero has noted that Var. 78 is the brightest known eclipsing binary of extreme Population II. One Algol type eclipser is in NGC 6838, and my tentative period is given for the first time in the Third Catalogue as 3.9 days. Perhaps the photometric study of this cluster by Arp and Hartwick will help determine whether or not it is a member. It is possible that our information on binary stars in globular clusters will increase substantially in the years ahead. The work of my colleague, Dr Christine Coutts, has shown that the binary character of an RR Lyrae variable can have a measurable effect on its observed period fluctuations.

Explosive variables account for half-a-dozen in globular clusters. Three U Gem stars have been observed, one near Messier 5, one near NGC 6712 and the other near Messier 30. They are all very faint, and little information is available on them. Perhaps they are all field stars. Three novae have been found, at least two of which seem to be members. The first was discovered visually in 1860 by Auwers in M80, and rose to just under naked eye visibility, magnitude 6.5 or so, changing the whole appearance of that compact little cluster. The second was found in 1949 by Mrs Margaret Mayall on spectrum plates of NGC 6553 taken at the Boyden Station of the Harvard Observatory in 1943. These plates cannot now be located for checking. The third, in Messier 14 was found by my collaborator Dr Amelia Wehlau in 1964 while working on plates I took with the 74-in. David Dunlap reflector in 1938. A series of 8 plates on nights in one week of June that year shows the star near 16th magnitude on all of them. It has never been found on any other plates of this or other observatories. It is very close to the centre of the cluster and there is no reason to think it is not an actual member. Most of the hunting for variables in globular clusters has not been of the sort that would detect novae. It was really good luck that those in NGC 6553 and M14 were found. Systematic novae search programs with plates taken bi-weekly over years might yield many more novae.

Now I would like to make a few remarks about the Canadian observational program which is actually the force behind these catalogues, and some previously unpublished material from it is included in the Third. My own work on globular clusters began in

1926 as a graduate student at the Harvard Observatory under Dr Harlow Shapley, and two people here today, Mrs Margaret Mayall and Miss Henrietta Swope were there then as graduate students also. In 1931 my husband Dr Frank Hogg was appointed to the staff of the Dominion Astrophysical Observatory, Victoria, B.C. In September, 1931 I began the cluster program there with the help of the Director, Dr J. S. Plaskett, and my husband, using the 72-in. reflector, at that time the second largest telescope in the world. I want to emphasize the extensive co-operation of many persons in this program for more than forty years. In 1935 my husband and I came to the David Dunlap Observatory and the 74-in. went into operation in May that year, at that time supplanting the DAO 72-in. as the second largest telescope. The program on variables in globular clusters has continued here ever since.

Great co-operation has come from the Observatory Directors, of whom the last two, Dr John F. Heard and Dr Donald A. MacRae are present today. National Research Council of Canada has given generous support to the program. Throughout all these years Gerry Longworth has kept the big telescope in top condition, and Frank Hawker and Anson Moorhouse have aided the program. My colleague, the late Professor Ruth Northcott gave valuable help, particularly during the hard war years.

Other, smaller telescopes have been used to supplement the 74-in. program. In 1939 the Director of the Steward Observatory of the University of Arizona, Dr E. F. Carpenter, gave me six weeks of time on the 36-in. there. Then the 19-in. built by Dr R. K. Young here has been used for many summers with student assistance. And the last five summers the new 16-in. atop this building, in the Burton Tower, has continued the program, run by Peter Chen, Kayll Lake, Rick Salmon and Chris Smith. The past year the new 24-in. of the University of Toronto has gone into operation on Las Campanas, Chile and superb cluster plates are coming back from it. Earlier, Dr Christine Coutts had obtained many plates for the program with the Michigan Schmidt on Cerro Tololo. As a doctoral student here Christine Coutts studied Messier 5, worked in Italy for a year under Dr Rosino, and is now continuing the program, specializing in period changes. Almost ten years ago Dr Amelia Wehlau of the University of Western Ontario began working on our plates, and now is acquiring her own with the new 48-in. at that University.

The catalogue has been completed with help from many astronomers who have sent in data, especially Dr B. V. Kukarkin, and from two librarians, Mrs Jean Lehmann and Mrs Sheila Smolkin, from our secretary Mrs Jennie Fabian and from my daughter Mrs Sally MacDonald who tabulated data.

The tables, figure and references accompanying this article will be included with the Third Catalogue, as *Publications of the David Dunlap Observatory* 3, No. 6.

DISCUSSION

Menzies: Have you considered extending your Catalogue to cover Magellanic Cloud globular clusters?

Hogg: Yes, but it has been difficult enough to complete this third Catalogue without including any extra material. Perhaps later.

Feast: Is it possible to say in how many clusters the search for RR Lyrae stars has been exhaustive (or essentially exhaustive)?

Hogg: It's very difficult to say. In Messier 10 which Bailey had hunted carefully, I was pleased to find two variables which he had missed. And then a few years later Dr Arp found an important variable which I had missed.

Dickens: Most variables are discoved by 'blinking' of plate pairs, which must lead to incompleteness in the discovery of small amplitude variables such as c-types, particularly in crowded areas.

Hogg: Yes, there is a selection effect working against the discovery of the small range c-type variables.

Wesselink: The absolute magnitudes of flare stars are so much fainter than those of RR Lyrae's that they are missed on plates taken for RR Lyrae's only, which accounts for the absence of known flare stars.

Buscombe: Real-time searches with image orthicon tubes (similar to the successful identification of supernovae in distant galaxies at Corralitos) could be productive for variables in globular clusters if attempted on large telescopes.

SOME PROBLEMS IN THE STUDY OF VARIABLE STARS IN GLOBULAR CLUSTERS

B. V. KUKARKIN

Astronomical Council of the U.S.S.R. Academy of Sciences, Moscow, U.S.S.R.

1. Introduction

Three quarters of a century separates us from the beginning of systematic observations of variable stars in globular clusters. During the last 45 yr the name of Mrs Prof. Dr H. B. Sawyer Hogg has always been in the astronomical literature. Her activity is closely connected with investigations of variable stars in globular clusters.

The complex dependences between the presence of variable stars of different types and the properties of the globular clusters were discovered long ago. For instance, the RR Lyr variables do not occur in clusters of high metallicity. Red variables on the contrary do not occur in clusters of very low metallicity. The presence and peculiarities of variable stars have been used for the classification of globular clusters (van Agt and Oosterhoff, 1959; Castellani *et al.*, 1970).

Unfortunately, little attention has been paid to the problem of the frequency of variable stars in globular clusters. Extremely interesting is the problem of a spatial distribution of variable stars relative to stars of different sequences in globular clusters. Also very important is the correct investigation of the period changes of the RR Lyr and W Vir variables in globular clusters. Especially interesting is the study of the period changes in connection with the position of variable stars on the colour-magnitude diagrams.

2. The Influence of Selection on the Frequency of Variable Stars in Globular Clusters

The main selection factor among the samples of different types of variables in globular clusters is the probability of discovery. The probability depends upon the type of variability, the range of variation, the apparent density of stars in clusters, etc. Thus the probability of discovery of RR_c-type stars is significantly smaller than for the RR_{ab}-type, while the probability of discovering eclipsing binaries is smaller than for the RR_c stars. The probability of finding U Geminorum type stars is still less. The usual method of observation of variable stars in globular clusters excludes the possibility of discovering U Gem stars and variables with very slow changes of brightness. The observers are usually interested in the RR Lyr variables and observe the globular clusters very intensively only during one or two months. The exposure time must be optimal for the RR Lyr variables. The probability of a U Gem variable flaring in this short interval is very small. Moreover, the absolute magnitude of U Gem variables at maximum is more than 2^m–3^m fainter than the RR Lyr stars at minimum. Therefore the U Gem variables in globular clusters are usually fainter than the limiting magnitude of the plates.

The majority of the red variable stars in globular clusters are very poorly studied. Many of them have not been discovered. Only a few observatories have photographs suitable for the discovery and study of the slowly varying red stars. Due to the initiative of Mrs H. B. Sawyer Hogg the David Dunlap Observatory has a unique collection of plates for this purpose.

Insufficient allowance for selection leads to erroneous conclusions. For example, the globular cluster NGC 3201 was investigated on the basis of many Harvard Observatory plates (Wright, 1941). 56 RR_{ab}-type stars and only one RR_c-type star were studied and some stars were not solved. Further observation (Kukarkin, 1971a, b) brought the number of RR_c variables to ten!

It is usually supposed that eclipsing binaries are rare in globular clusters. Actually the population of globular clusters does not correspond to the population of the majority of eclipsing binaries in our Galaxy or the Magellanic Clouds, but it is known that in globular clusters there exist stars of the U Gem-type and Novae, and in the extremely old open clusters there are W UMa-type stars. All these stars are much fainter than the RR Lyr variables. Unfortunately no one has systematically searched for variable stars in this magnitude interval in globular clusters.

It is clear that customary ideas on the frequency of variable stars of different types will be changed after accurate consideration of selection.

Thus the concept of frequency demands a more accurate definition. The frequency must be determined not by the absolute number of objects, but by the ratio of this number to the number related with the richness of the stellar system studied (e.g. the absolute magnitude M, the number of stars brighter than a set absolute magnitude, etc.).

The problem of the frequency of variable stars and other objects was considered some years ago (Kukarkin, 1968).

The frequency of given types of objects in a given stellar system is the ratio of the absolute number of objects to the total number of stars or to some value proportional to this number. The influence of selection and life-time must be considered.

3. The Frequency of RR Lyr-, W Vir-, and RV Tau-Type Stars

In the globular cluster ω Cen 139 RR Lyr-, 6 W Vir- and one RV Tau-type stars are known. In the globular cluster NGC 5053 only 10 RR Lyr stars and no W Virginis stars are known. According to the absolute number of variable stars ω Cen is extremely rich. But according to the richness of these two clusters this conclusion is erroneous. The absolute magnitude M_V of ω Cen is $-10\overset{m}{.}2$ and of NGC 5053 only $-6\overset{m}{.}1$ (Kukarkin and Russev, 1972). We assume that the general number of stars is proportional to the luminosity. We then find that ω Cen is 40 times richer than NGC 5053 so that it is necessary to reduce the number of variables in ω Cen by the factor 40. We obtain $139/40 \simeq 3$–4 and $7/40 \simeq 0$. Consequently the frequency of RR Lyr stars in NGC 5053 is 3–4 times greater than in ω Cen!

In Table I the frequencies of RR Lyr stars are given for well observed globular clusters with reliable absolute integral magnitudes. In the first column is given the designation of cluster according to NGC Catalogue; in the second – the observed

TABLE I

The frequencies of RR Lyrae variables in globular clusters

NGC	n	M_V	N	IM	
104	2	−8.93	0.5	0.56	4
362	12	−7.09	17	0.44	3
2419	25	−8.30	12	0.34	1
3201	83	−6.67	178	0.41	3
4147	16	−5.91	69	0.33	3
4590	37	−6.9:	64	0.22	2
4833	11	−7.47	11	0.31	3
5024	33	−8.54	12	0.30	4
5053	10	−5.91	43	0.25	3
5139	137	−10.19	11	0.35	5
5272	200	−8.29	97	0.38	5
5466	20	−6.53	49	0.26	4
5824	25	−7.61	22	0.29	2
5897	5	−7.11	7	0.26	1
5904	92	−8.24	46	0.39	5
6121	40	−6.24	127	0.45	4
6171	22	−6.27	69	0.50	5
6205	2	−8.05	1	0.34	4
6218	0	−7.15	0	0.35	4
6229	20	−7.26	25	0.42	3
6254	0	−7.26	0	0.36	4
6266	81	−8.69	26	0.45	2
6333	11	−7.76	9	0.32	1
6341	13	−8.01	8	0.25	4
6352	0	−5.31	0	0.65	1
6362	23	−6.18	78	0.43	3
6397	0	−6.32	0	0.33	4
6402	62	−8.79	19	0.42	4
6426	12	−5.63	67	0.33:	1
6626	7	−7.99	4.5	0.47	1
6637	0	−7.84	0	0.66	2
6656	17	−8.17	9	0.32	4
6712	9	−6.81	17	0.51	3
6715	54	−8.61	19	0.43	3
6723	19	−7.38	21	0.51	3
6752	0	−7.35	0	0.35	1
6779	0	−7.10	0	0.31	4
6809	6	−7.6:	5:	0.32	1
6838	0	−4.90	0	0.59	4
6934	44	−6.91	76	0.39	3
6981	33	−6.70	69	0.38	4
7006	60	−7.27	74	0.37	3
7078	78	−8.84	23	0.25	5
7089	17	−8.92	5	0.30	4
7099	3:	−6.98	5:	0.26	1
7492	3	−4.89	30	0.30	1

number n of RR Lyr stars; in the third – the absolute magnitude of clusters M_V; in the fourth – the number N of RR Lyr stars reduced to $M_V = -7\overset{m}{.}5$; in the fifth – the index of metallicity IM and in the sixth – the estimate of reliability (5 – very good, 1 – very poor).

It is clearly seen that there are three areas of frequency of RR Lyr stars in globular clusters.

(I) In all the clusters with $IM \geqslant 0.57$ the RR Lyr stars are completely absent.

(II) In the clusters with $0.57 > IM > 0.32$ the number of RR Lyr stars increase with decreasing values of IM. The clusters NGC 3201, 6121 and 5272 have very high frequencies of RR Lyr stars. (The cluster IC 4499 is probably also extremely rich but unfortunately the type of variable star in this cluster is unknown and the value of M_V is very unreliable). The values of IM in the second area are overlapped by the values in the third area, but both are very well separated (see Figure 1).

(III) Some clusters with $IM < 0.37$ have no RR Lyr stars (NGC 6218, 6254, 6397 etc.). With the decreasing of IM the number of RR Lyr stars increases.

The globular clusters which fall in the second area of Figure 1 show a preferential concentration of RR_{ab} variables, particularly those of the short-period group; clusters with longer-period RR_{ab} stars tend to lie in the third area, as do those with RR_c variables. Clusters with W Vir stars are present in both the second and third areas but absent from the first. Clusters with red variables are present in all three areas, but those with Mira stars are found only in the first area.

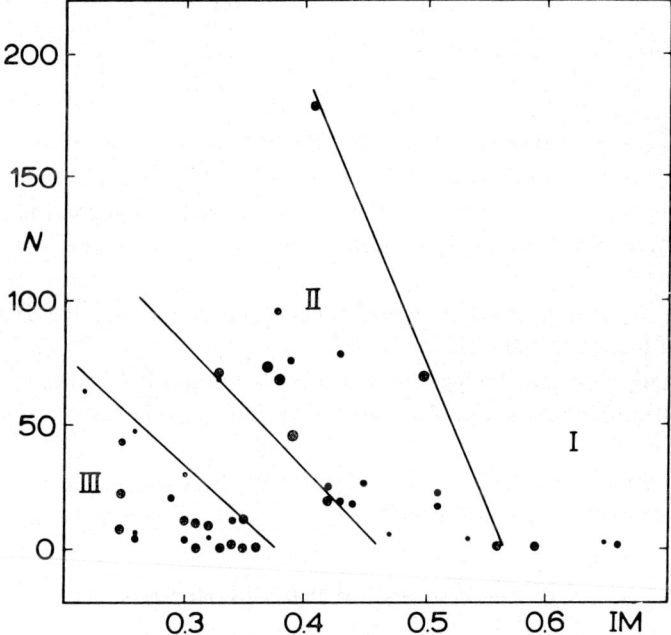

Fig. 1. The frequency of RR Lyrae variables as a function of the index of metallicity of the globular clusters.

It is clear that the two-dimensional $N-IM$ relation, while requiring interpretation, is insufficient for describing the properties of globular clusters.

The following clusters have a very low frequency of RR Lyr stars: NGC 6397, 6218, 6779, 6254. Perhaps the high frequency of RR Lyr variables is connected with the index of metallicity between 0.4 and 0.5. The preliminary diagram of frequency of the RR Lyrae stars as a function of metallicity of the globular clusters is given in Figure 1.

The true frequencies of the RR Lyr stars have already been used for the classification of globular clusters (Mironov, 1973), but the problem deserves a thorough study.

TABLE II

W Vir — RV Tau variables and integrated properties of globular clusters

NGC	n	M_V	N	IM
5024	1	−8.54	0.3	0.30
5139	7	−10.22	0.6	0.35
5272	1	−8.32	0.5	0.38
5904	2	−8.28	1.0	0.39
6093	1	−8.22	0.5	0.36
6205	3	−8.04	1.8	0.34
6218	1	−7.19	1.3	0.35
6254	2	−7.24	2.5	0.36
6402	5	−8.75	1.6	0.42
6656	1	−8.13	0.6	0.32
6779	1	−7.05	1.5	0.31
7078	1	−8.84	0.3	0.25
7089	4	−8.94	0.9	0.30

In Table II the frequencies of W Vir plus RV Tau stars are given. It is necessary to remember the very small number of W Vir and RV Tau stars and the possibility of large fluctuations. But it is very probable that in the well studied clusters practically all the W Vir and RV Tau stars have been discovered. The structure of Table II is the same as Table I.

The following clusters have a high frequency of W Vir and RV Tau variables: NGC 6254, 6205, 6402 and 6779.

It was recently shown that a high frequency of W Vir and RV Tau stars is connected with a high helium abundance (Mironov, 1973), but this conclusion requires further investigation.

The recent start on systematic observations of W Vir and RV Tau stars in globular clusters (Demers, 1971: Kukarkin and Rastorguev, 1972) is very welcome.

4. The Frequency of Red Variable Stars

The red variable stars (excluding Mira-type stars) are encountered more frequently in globular clusters than W Vir and RV Tau stars. A special list of red and similar variables in globular clusters was compiled by Stothers (1963). For some red variables the spectra and radial velocities were obtained by Joy (1949), but the photometric study of these stars is unsatisfactory. The principal reason for this is the very short time intervals during which most observations are obtained.

The great difficulty in the photometry of red variables is the unacceptability of the usual *UBV* system. This unacceptability is caused by the very strong TiO absorption bands, especially in the *V* region. Therefore the position of the red stars on the $V-(B-V)$ diagrams does not show the true position on the luminosity-temperature diagram. The search for good substitutes for the *UBV* system is extremely significant (see Eggen, 1968, 1969; Lloyd Evans, 1971; Lockwood and Wing, 1971; Russev, 1972). It is obvious that further observations must be made without the use of the *V* region; observations in the *B* region are preferable. Therefore the observations in regions similar to *B* may be very useful in the study of the variability of the red stars.

More than 15 yr ago it was noted that the instability of red giants in globular clusters increases with their luminosity (Walker, 1955), an effect recently confirmed by Russev (1971). Figure 2 demonstrates the advantage of the photometric region *B* as compared

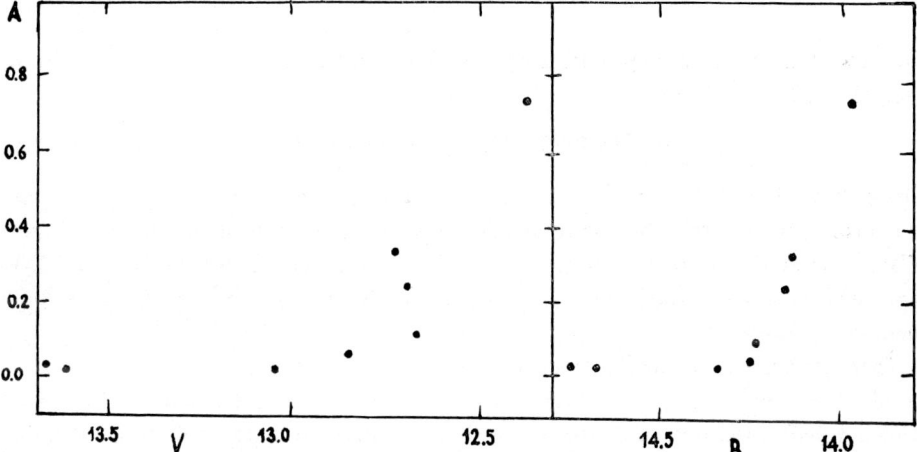

Fig. 2. The increase of instability with luminosity in the photometric region *V* (left) and *B* (right). The advantage of the *B* region is evident.

with the *V* region. The left part of the figure gives the 'magnitude *V* – Amplitude' relation (Russev, 1971). On the right part of the figure the *V* magnitudes are replaced by the *B* magnitudes. When the colour-magnitude diagram is expressed in a photometric system better than the $V, (B-V)$ one, all the red variables in globular clusters fall to the extreme right end of the giant branch. It is possible that all these stars are at the turning-point of evolution, near the helium flash.

The presence of red variables in such different globular clusters as 47 Tuc, M71, M13 and ω Cen is very significant. It is very probable that in a suitable photometric system the red variables would be separated. This separation should be related to age and chemical composition.

We now have in Moscow a 12-yr series of B observations of such different clusters as M56 (index of metallicity $IM = 0.30$) and M71 ($IM = 0.60$).

In Table III the frequencies of the red variables are given. The structure of Table III is the same as in Table I. The numbers of red variables are small and there is no certainty that all the red variables have been discovered.

TABLE III

Red variables and integrated properties of globular clusters

NGC	n	M_v	N	IM
104	6	−8.93	1.6	0.56
5139	10	−10.22	0.9	0.35
5272	3	−8.32	1.4	0.38
6121	2	−6.20	6.6	0.45
6205	2	−8.04	1.8	0.34
6656	3	−8.13	1.7	0.32
6779	4	−7.05	6.1	0.31
6838	2	−4.91	21.7	0.59

The new data on red variable stars are given by Lloyd Evans and Menzies in this book (pp. 151–163).

5. The Frequency of Mira-Type Stars

The membership of the three Mira-type stars in the globular cluster 47 Tuc is certain. It is also very probable that there are two Mira-type stars in the cluster NGC 6637 (Catchpole *et al.*, 1970). The index of metallicity IM of 47 Tuc is 0.57, of NGC 6637 0.62, which makes it likely that the presence of Mira-type stars is typical of high metallicity clusters.

There are some globular clusters with one or two Mira-type stars in their neighbourhoods. For some clusters (e.g. NGC 6093, 6171, 6656) it is clear that the Mira-type stars are field stars, but we do not know this for others. It is desirable that the radial velocities of the possible Mira-type members of globular clusters be determined. Some new data are given in Feast review (see pp. 131–144).

6. The Frequency of Eclipsing Binaries

Some globular clusters (e.g. NGC 3201, ω Cen, 6838) have one eclipsing binary. There is no convincing evidence for their membership, however, although in some cases their apparent magnitude does not rule out membership. But it is too early to speak about the frequency of eclipsing binaries.

The problem of the U Gem stars and novae in globular clusters was considered recently (Kurochkin and Kukarkin, 1966; Kukarkin and Mironov, 1970). A reliable solution of the problem demands systematic searches for variable stars in the magnitude interval usually neglected by observers (from 2^m to 8^m fainter than RR Lyr stars).

Arp has kindly given us some plates of the globular cluster M5 made with the 200-in. Palomar telescope (limiting magnitude about 21^m–22^m). Unfortunately only one pair of plates was available for comparison. Kurochkin made a thorough comparison of this pair, but no eclipsing or other variables except V 101 were discovered. However, this is not unexpected because of the very small probability of discovering eclipsing binaries.

The comparison of two pairs of plates taken with the 40-in. Burakan Schmidt telescope (limiting magnitude about 19^m–20^m) has revealed one faint variable in the

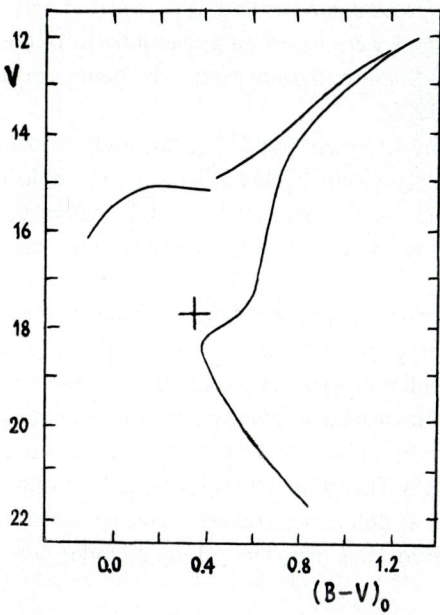

Fig. 3. The position on the colour-magnitude diagram of a new faint variable star in the globular cluster in M92.

neighbourhood of the globular cluster M92. This star is situated on the colour-magnitude diagram near the turn-off point of the main sequence. Its position is very similar to the position of W UMa variables (see Figure 3), but unfortunately the small number of plates does not allow the period or light curve to be determined.

The problem of eclipsing binaries in globular clusters requires very frequent observations.

Special attention must be paid to the search for eclipsing effects among RR Lyr stars. The discovery of such a star in the Ursa Minor dwarf galaxy (Kholopov, 1970, 1971) is very promising.

It is desirable to investigate thoroughly all pronounced fadings of RR Lyr stars in globular clusters, especially at light minimum. It is probable that such fadings have been accepted by observers as due to observational errors.

7. The Dissipation of Variable Stars and the Dynamical Evolution of Globular Clusters

The dynamical history of any autonomic stellar system inevitably involves the problem of dissipation. The construction of colour-magnitude diagrams for comparatively outlying regions of globular clusters has shown that the number of cluster members is still high. Many papers by Kholopov have demonstrated the presence of very extended coronae (see e.g. Kholopov, 1968a).

The presence of appreciable members of RR Lyr stars in far outlying regions of some globular clusters was demonstrated very convincingly by Kurochkin (1961, 1962). These investigations were based on a search for variable stars up to 5° around some globular clusters. Similar investigations are being carried out by the Italian astronomers (Rosino, 1970).

The RR Lyr stars are extremely useful for the study of dissipation, because their relationship with the given globular cluster is easy to establish. The probability that the apparent magnitude, period, shape of the light curve, etc. of the field RR Lyr variables will be similar to those in the given globular cluster is so slight that no doubt will arise.

In recent years many articles on the dynamic evolution of spherically symmetric stellar systems have been published (see for example King, 1962, 1966; Larson, 1970). Efficiently organised studies of variable stars in the extended surroundings of globular clusters must yield rich material for investigating the dynamical evolution of clusters. It is also possible that this material may give an independent method for determining the masses of globular clusters. Therefore it is most desirable to study many globular clusters of different masses and at different distances from the centre of our Galaxy. It would also be interesting to study this problem among globular clusters in the Magellanic Clouds.

8. The Distribution of Stars of Different Types in Globular Clusters and the Problems of Stellar Evolution

The problem of the distribution of stars of different types in clusters was studied long ago. But these investigations were not always correct. For instance Woolf's attempt to estimate the mass ratio of stars in the red and blue parts of the horizontal branch was hardly serious (Woolf, 1964), being based on a simple count of star numbers at different distances from the centre of the globular cluster M3.

The estimate of the mass ratio of RR Lyr and W Vir stars by Kukarkin and Voroshilov (1971) must be considered as preliminary. This estimate was based on the determination of the density gradient of the RR Lyr and W Vir stars. But the number of W Vir stars in globular clusters is small and it was therefore necessary to construct a synthetic globular cluster by reducing individual clusters to a common distance and radius. However, the established fact of the absence of W Vir stars and the presence of RR Lyr stars at great distances from the centres of clusters gives evidence of the larger masses of W Vir stars.

In the past ten years more correct investigations have been carried out (see for example Kholopov, 1968a, b; Blaghikh and Castellani, 1971; Castellani *et al.* 1972). These papers showed that it is impossible to accept the equality of masses of all stars on the giant or horizontal branches. For the RR Lyr stars Kholopov has derived results contradicting the mass estimates for stars of other types. This contradiction is probably connected not only with the peculiarities of the RR Lyr stars themselves, but also with the space density of the stellar medium.

The correct study of the problem of space distribution of stars of different types in globular clusters is extremely important. The determination of the mass ratio of stars of different types must be the test of the numerous theoretical calculations (see for example Schwarzschild, 1970; Schwarzschild and Härm, 1970; Rood, 1970; Iben and Rood, 1970; Demarque and Geisler, 1971; Iben, 1972; Demarque and Mengel, 1972; etc.).

9. The Variability of the Periods of Variable Stars

In the twenties and thirties of our century it was customary to represent the deviations of the epochs of variable stars from the linear elements by second order equations. The coefficient of the second order term was accepted as the value describing the progressive change in the period. In this way it was established that the instability of the period of the classical Cepheids increases with the period itself (Florja and Kukarkin 1932). This same technique was applied to the RR Lyr stars in globular clusters (see for example Martin, 1938, 1942). The coefficient of the second order term was designated in many articles as β. Such an approach to the problem at that time is excusable because the second order equations represented the deviations of the epochs satisfactorily. But single variable stars show more complex deviations.

Numerous new observations and a considerable increase in the intervals covered by observations have shown that in reality the phenomenon is very complex. The periodicity of a pulsating star undergoes random fluctuations which are spontaneous variations having quite different origins. We often observe successive changes of period, in which it increases then decreases, and vice versa. In these cases the epochs can be represented very well by two or more systems of linear equations. This was clearly demonstrated as long ago as the mid-fifties by Parenago (1956).

The intervals of observation of the RR Lyr stars are short and reach at best 80 yr, yet it is clear that the observed changes of period must be considered as noise. This noise may in fact veil the slow evolutionary changes.

It is not impossible that a proper examination of the period-noise of many RR Lyr stars in some rich globular clusters by modern methods of mathematical statistics may lead to the discovery of the evolutionary changes of periods. But it is necessary to proceed with extreme caution. The level of noise varies among different types of RR Lyr stars in the same globular cluster. It is extremely large for the RR Lyr stars showing the Blazhko-effect, but there are other RR Lyr stars with quite constant periods over intervals of 70 or even 80 yr.

The quantity β is also a certain measure of noise, but the approach to the problem

Dickens: What are typical evolutionary times between horizontal branch and W Virginis phases? Relaxation times could be as short as 10^8 yr near the centres of some globular clusters, although for large clusters such as M3 they are an order of magnitude longer.

Wesselink: No δ Scuti stars were found on long exposure plates of ω Cen by the Leiden observers at the Radcliffe Observatory.

Graham: Are there not special observational problems in detecting variable stars of faint absolute magnitudes in globular clusters?

Wesselink: The problem is of course more difficult than the discovery of RR Lyrae's for two reasons:

(1) The fainter brightness requires longer exposures.

(2) The absolutely fainter stars exhibit changes which are more rapid.

Nevertheless the sensitivity of presently existing equipment is good enough to find such stars if they exist.

Rosino: It would be very difficult to find flare stars in globular clusters, since these stars have an extremely rapid variation, the time from minimum to maximum and back to minimum being sometimes as short as 5–10 min – so if the exposure time is of the order of 30^m these stars will certainly be lost.

OBSERVATIONS OF FAINT GLOBULAR CLUSTERS

L. ROSINO and G. PINTO
Astrophysical Observatory, Asiago, Italy

The present communication reports some results of observations made at Asiago on three faint globular clusters discovered on Palomar Sky Survey plates (Abell 1955).

Palomar 2 ($4^h43^m.1$, $+31°23'$, 1950; $l=170°$, $b=-9°$). This cluster was studied some years ago by McCarthy and Treanor (1964) on infrared plates taken with the Schmidt telescope of the Vatican Observatory. After an extensive discussion the authors concluded that the object was either a peculiar globular cluster at an estimated distance of 63 kpc or an old galactic cluster at a distance of the order of 16 kpc.

In 1957–58 a series of infrared photographs (IN hypersens. + RG5) were obtained at Asiago at the newtonian focus of the 122 cm telescope, and one plate was taken in 1969 with an RCA-Carnegie S1 image tube through an RG8 filter. Some blue and yellow (103a-D+GG11) plates were also available.

The cluster is scarcely visible on the blue plates; it becomes more clear in yellow and quite sharp in infrared light (Figure 1). Its structure, on the infrared plates, is that of a globular cluster with a moderate concentration. Comparison with other distant underexposed globular clusters shows that the general structure is the same. The apparent diameter has been estimated to be about 3′, but the cluster would certainly become larger if the faintest components, weakened by strong interstellar absorption, and barely visible in the infrared, should emerge. More than 150 stars have been counted on the Asiago infrared plates within 90″ of the centre of the cluster, of which at least 20 are 1.5 mag. brighter than the plate limit.

Assuming tentatively a linear diameter between 20 and 40 pc, which seems reasonable for a globular cluster of intermediate size, with an apparent diameter of 3′ a distance of between 23 and 46 kpc is obtained, corresponding to a median distance modulus of about 17.4. Considering the deep absorption in the direction of the cluster, at least 2 mag. in the blue, it is very unlikely that the horizontal branch can be reached with the Asiago telescope, even in the infrared.

The cluster has been carefully examined at the blink for the finding of variables, but the result has been negative: no variables have been found. Since the plates cover a period longer than one year, it is likely that the cluster does not contain red variables of the semiregular or long period type. However, the possibility that RR Lyrae variables may be present cannot be ruled out.

Palomar 4 ($11^h27^m.1$, $+29°12'$, 1950; $l=202°$, $b=+72°$). Two slow variables have been discovered in this extremely faint cluster by Rosino (1957) and by Burbidge and Sandage (1958), who determined the *c-m* diagram and remarked on its peculiarities. The horizontal branch is almost absent, being reduced to a short red stub, although

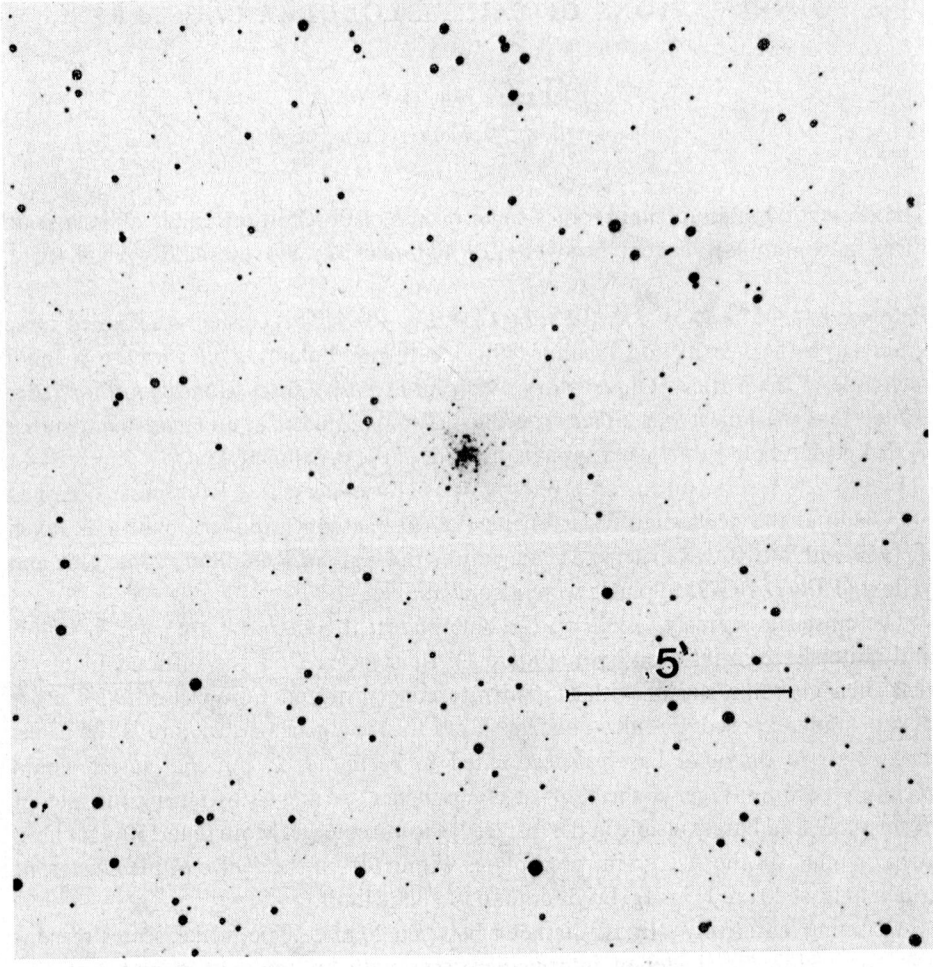

Fig. 1. The globular cluster Pal. 2. Infrared IN hypers. +RG5, 122 cm tel. north at the top.

the ΔV at $B-V=1.4$ is about 2.7. The two red variables are located at the top of the giant branch.

In his previous paper Rosino estimated that the periods of the variables were about one-hundred days. Burbidge and Sandage gave periods of $131^{d}\!.6$ for Var. 1 (their star No. 44) and $166^{d}\!.7$ for Var. 2 (star No. 25).

The present observations are reported in Table I. The magnitudes of the variables have been estimated on fourteen blue plates obtained between 1963 and 1967 (one in 1972) with the 122 cm telescope, and from some of the best blue photographs taken with the 67 cm Schmidt in 1971–72. Fourteen infrared Schmidt photographs (IN hypersens. +RG5) were also available.

In the reduction of the blue material (103a-O without filter) the same comparison sequence has been used as in the previous paper. Eye estimates of the magnitudes have

TABLE I
Magnitudes of variables No. 1 and 2 in Palomar 4

Date	UT	JD	No. 1	No. 2	
1963 Feb 26	23^{h}08^m	243 8087	19.5	19.0	Newt.
Mar 24	22 20	8113	19.55	19.35	
1964 Feb 12	23 58	8438	18.3	19.3	
Mar 7	22 52	8462	–	17.95	
Mar 9	22 36	8464	19.7	17.9	
1965 Apr 1	00 55	8852	19.2	18.7	
1966 Jan 20	22 15	9146	19.6	18.6	
Mar 2	3 25	9187	17.8	19.0	
Mar 14	21 55	9199	17.75	19.35	
Mar 21	2 39	9206	17.85	19.3	
1967 Feb 16	1 18	9538	19.1	19.1	
Mar 13	22 38	9563	18.05	18.2	
Mar 15	22 58	9565	17.95	18.15	
1971 Mar 2	00 02	4 1013	17.5:	–	Schmidt
Mar 3	00 25	1014	17.9:	–	Schmidt
1972 Feb 12	1 44	1360	18.8:	19.2:	Schmidt
Mar 20	0 16	1397	17.65	18.1	Newt.
May 2	20 36	1440	–	18.15:	Schmidt

been made independently by Bianchini and Rosino. The mean value of the estimated magnitudes is reported in Table I. The elements and light curves have been determined by Pinto with an electronic computer. The elements of the two variables are:

Var. 1: $T =$ JD 243 5922 $P = 130^d.50$
    2:                  5938   109.30.

The light curves are shown in Figure 2. The elements of Var. 1 also satisfy the observations of Burbidge and Sandage, after a correction of -0.5 is made to their magnitudes, which takes into account the different system of comparison stars. In the light curve the observations of Burbidge and Sandage are marked with open circles.

The elements of Var. 2 represent fairly well the old and new observations of Asiago, but some of the magnitudes given by Burbidge and Sandage do not fit the mean light curve.

The dispersion observed in the light curves may be partly due to errors of observation, the variables being very weak on the Asiago material, and partly to the fact that the two stars are semiregular and do not repeat the same light curve from cycle to cycle.

As shown in Figure 2, Var. 1 has an amplitude of $2^m.2$ pg between 17.6 and 19.8 pg (mean values); Var. 2 has amplitude of $1^m.8$ between 17.6 and 19.4 pg. Median pg magnitudes: 18.7 and 18.5. The variables have also been estimated on the Schmidt infrared plates. They are the brightest components of the cluster, even at minimum. Var. 1 has an infrared amplitude of about $0^m.9$; Var. 2: $0^m.85$. The $B-I$ color index increases from maximum to minimum.

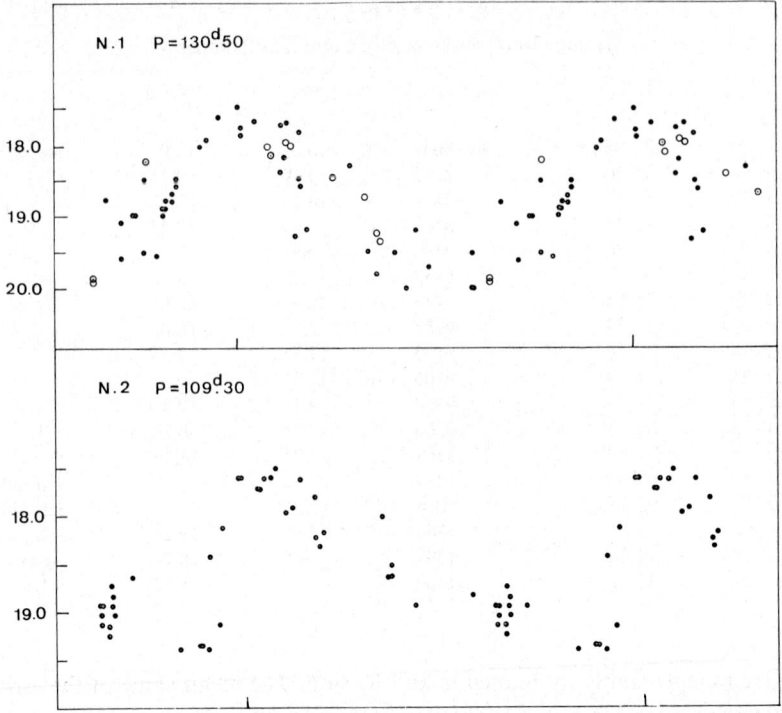

Fig. 2. Light curves of Var. No.1 and No.2; circles indicate Palomar observations corrected by −0.5.

With a mean median pg absolute magnitude −2.4 (if a distance modulus of $m - M = 21$ is adopted for the cluster) and periods between 100 and 150 days, the two variables can be classified as red semiregulars, a type which is not uncommon in globular clusters with a relatively high metal abundance.

Palomar 5 (15^h13^m5, $+0°5'$, 1900; $l = 1°$, $b = 46°$). This giant-poor cluster was discovered by Baade in 1950 on Palomar Schmidt plates. Five RR Lyrae variables were later found by Rosino (1951), and their periods determined by Pietra (1956) and Mannino (1956). The elements have been slightly improved by Kinman and Rosino (1962) as follows:

Var. 1	$P = \overset{d}{.}0293230$	Ampl. 0.42 B
2	.332467	0.40
3	.329953	0.40
4	.286362	0.48
5	.252395	0.30.

The peculiarity of this cluster, besides its looseness and the scarcity of giant stars, is the fact that it contains five RR_c Lyr variables without any RR_{ab} stars, a case which is *unique* among globular clusters.

Other photographs of the cluster have been taken at Asiago since 1962. The magnitudes determined by examing the new material fit perfectly the light curves obtained with the preceding elements. However, a tentative attempt has been made by one of us (G.P.) to see whether some alternative period, larger than $0^d.6$, might fit all the observations. The result has been completely negative. It is therefore confirmed that the five RR Lyrae variables of this cluster are all of type c.

References

Abell, G. O.: 1955, *Publ. Astron. Soc. Pacific* **67**, 258.
Burbidge, E. M. and Sandage, A.: 1958, *Astrophys. J.* **127**, 527.
Kinman, T. D. and Rosino, L.: 1962, *Publ. Astron. Soc. Pacific* **74**, 499.
Mannino, G.: 1956, *Pubbl. Oss. astr. Univ. Bologna* **VI**, 17.
McCarthy, M. F. and Treanor, P. J.: 1964, *Ric. astr. Specola Vatic.* **6**, 24.
Pietra, S.: 1956, *Pubbl. Oss. astr. Univ. Bologna* **VI**, 16.
Rosino, L.: 1951, *Pubbl. Oss. astr. Univ. Bologna* **V**, 15.
Rosino, L.: 1957, *Contr. Oss. astrofis. Univ. Padova* No.85.

PLANETARY NEBULAE, GLOBULAR CLUSTERS
AND THE GALACTIC HALO

SIDNEY VAN DEN BERGH

David Dunlap Observatory, University of Toronto, Richmond Hill, Ontario, Canada

Abstract. Planetary nebulae are used as a tracer for halo Population II stars. A comparison of the number of planetary nebulae in the galactic pole caps ($|b| \geqslant 45°$) with the number of planetaries in globular clusters suggests that (to within a factor of 2 or 3) the galactic halo has a luminosity $L_B \simeq 2 \times 10^8\, L_\odot$. From the work of Oort it is estimated that the galactic halo has a mass of *at least* $6 \times 10^9 \mathfrak{M}_\odot$ so that $\mathfrak{M}/L > 30$ for the galactic halo.

Planetary nebulae are generally regarded as members of the old disk population (O'Connell, 1958) of the Galaxy. Nevertheless there are some planetary nebulae which are undoubtedly members of a metal-poor halo population. The best known examples of this type of object are the planetary nebulae K 648 in M15 (Küstner, 1921; Pease, 1928; O'Dell *et al.*, 1964; Peimbert, 1973) and the high-velocity halo planetary 49 + 88°1, which has been shown to be metal-poor by Miller (1969).

It is the purpose of this paper to point out that high-latitude planetary nebulae, for which observational data should be quite complete, can be used as a tracer for halo Population II. For the purpose of the present investigation we shall define halo Population II as that population component of the Galaxy which has a space distribution similar to that of the cloud of globular clusters in which the Galaxy is embedded.

According to Arp (1965) the Galaxy contains 119 globular clusters of which 16 (13 percent) are located in the northern ($b \geqslant +45°$) and southern ($b \leqslant -45°$) pole caps. A total of 8 planetary nebulae (Perek and Kohutek, 1967) are known within this same area; four in the north pole cap and four in the south pole cap. This sample of planetary nebulae at high latitudes should be virtually complete because this area has been thoroughly searched with powerful instruments such as the Palomar 48-in. Schmidt (Abell, 1966).

Of the 8 known planetary nebulae with $|b| \geqslant 45°$ at least three (NGC 246, NGC 1360 and NGC 3587) are relatively nearby objects within 600 pc of the Sun (Cahn and Kaler, 1971), which are probably members of the galactic disk population. The total number of true halo-type high-latitude ($|b| \geqslant 45°$) planetaries is therefore $\simeq 5$, Assuming such objects to have a space distribution similar to that of globular clusters then yields a total population of $5 \times (119/16) = 37$ such objects in the entire Galaxy.

So far K 648 in M15 is the only planetary that is known to be located in a globular cluster. A search for additional planetaries in globular clusters by Feibelman (unpublished) and by Andrews (Thackeray, 1971) has so far remained fruitless. In what follows it will, more or less arbitrarily, be assumed that the total number of planetary nebulae in *all* galactic globular clusters is three. The total population of bright evolved stars in the halo is therefore very approximately $(37/3) \simeq 12$ times greater than that in all galactic globular clusters together. The total (blue) luminosity of all galactic globular clusters is $\simeq 1.7 \times 10^7 \, L_\odot$ so that the total luminosity of the halo component of the Galaxy becomes $\simeq 2 \times 10^8 \, L_\odot$. (This value is quite insensitive to the luminosity function of faint K and M dwarfs which might differ in globular clusters and in the halo).

Oort (1965) has presented powerful arguments which suggest that at least 5 percent of the mass of the Galaxy is contributed by faint halo subdwarfs of spectral types K and M. Adopting a total galaxy mass of $1.3 \times 10^{11} \mathfrak{M}_\odot$ (Innanen, 1966) then yields a mass $\geqslant 6.5 \times 10^9 \mathfrak{M}_\odot$ for the halo. Comparison of this value with the halo luminosity obtained above yields $\mathfrak{M}/L > 30$ for the galactic halo. This is very much larger that the value $\mathfrak{M}/L \simeq 0.5$ that is obtained for galactic globular clusters (Schwarzschild and Bernstein, 1955; Feast and Thackeray, 1960). The fact that $\mathfrak{M}/L > 30$ for the halo and $\mathfrak{M}/L \simeq 0.5$ for globular clusters militates against the hypothesis (Peebles and Dicke, 1968; Peebles, 1969) that the halo was formed from gas clouds in which physical conditions were similar to those prevailing in proto-globular clusters.

The high $\mathfrak{M}/L$ ratio that is found in the galactic halo might be related to the high $\mathfrak{M}/L$ ratios that have been observed in the outer regions of M31 and M33. In M33 Boulesteix and Monnet (1970) find that the mass-to-light ratio changes from $\mathfrak{M}/L \simeq 2.5$ at $\varpi = 5'$ to $\mathfrak{M}/L \simeq 20$ at $\varpi = 40'$. A similar trend in the mass-to-light ratios in M31 has been found by Gottesman and Davies, (1970).

According to van den Bergh (1972) the Galaxy has a total luminosity $L_B = (1.1 \pm 0.3) \times 10^{10} \, L_\odot$. Comparison with the halo luminosity $L_B \simeq 2 \times 10^8 \, L_\odot$ that was obtained above indicates that ~ 2 percent of the luminosity of the Galaxy is contributed by metal-poor halo stars.

References

Abell, G. O.: 1966, *Astrophys. J.* **144**, 259.
Arp, H. C.: 1965, Stars and Stellar Systems **5**, 401
Bergh, S. van den: 1972, *Astron. Astrophys.* (in press).
Boulesteix, J. and Monnet, G.: 1970, *Astron. Astrophys.* **9**, 350.
Cahn, J. H. and Kaler J. B.: 1971, *Astrophys. J. Suppl.* **22**, 319, (No. 189).
Feast, M. W. and Thackeray, A. D.: 1960, *Monthly Notices Roy. Astron. Soc.* **120**, 463.
Gottesman, S. T. and Davies, R. D.: 1970, *Monthly Notices Roy. Astron. Soc.* **149**, 263.
Innanen, K. A.: 1966, *Z. Astrophys.* **64**, 158.
Küstner, F.: 1921, *Veröff. astro. Inst. Univ. Bonn* **15**, 1.
Miller, J. S.: 1969, *Astrophys. J.* **157**, 1215.
O'Connell, D. J. K.: 1958, *Stellar Populations*, North Holland Pub. Co., Amsterdam, p. 533.
O'Dell, C. R., Peimbert, M., and Kinman, T. D.: 1964, *Astrophys. J.* **140**, 119.
Oort, J. H.: 1965, *IAU Trans.* **XIIA**, 789.
Pease, F. G.: 1928, *Publ. Astron. Soc. Pacific* **40**, 342.
Peebles, P. J. E.: 1969, *Astrophys. J.* **157**, 1075.
Peebles, P. J. E. and Dickie, R. H.: 1968, *Astrophys. J.* **154**, 891
Peimbert, M.: Paper presented at the 18th Liège Symposium June 26–28, 1972.
Perek, L. and Kohoutek, L.: 1967, *Catalogue of Galactic Planetary Nebulae*, Academia Publ. House, Prague.
Schwarzschild, M. and Bernstein S.: 1955, *Astrophys. J.* **122**, 200.
Thackeray, A. D.: 1971, *Quart. J. Roy. Astron. Soc.* **12**, 320.

DISCUSSION

Feast: (1) Griffin's radial velocities will possibly increase $\mathfrak{M}/L$ for globular clusters by a moderately large factor. (2) It is important in your calculations that the ratio of planetaries in the region of galactic centre to globular clusters there is much greater than in the halo.

van den Bergh: The observations by Wilson and Coffeen (*Astrophys. J.* **119**, 197, 1954) and by Griffin (*Observatory* **92**, 29, 1972) refer to different clusters. For M92 Wilson and Coffeen obtain a radial velocity dispersion of 4.4. km s^{-1} compared to Griffin's value of 6 km s^{-1} in M13. This small difference in velocity dispersion may, at least in part, be due to the fact that M13 is a more compact cluster (Arp, *Stars and Stellar Systems* **5**, 401, 1965) than is M92. In summary there is as yet no indication that the published $\mathfrak{M}/L$ ratios in globular clusters need to be revised significantly.

(2) The nuclear bulge of the Galaxy (van den Bergh, *Publ. Astron. Soc. Pacific* **84**, 306, 1972) contains a dominant old disk population in which a few halo population stars are embedded. The vast majority of the planetary nebulae in the direction of the galactic centre should therefore belong to the old disk population.

Demarque: Would Dr Schwarzschild care to comment on a recent study done at Princeton by Ostriker and Peebles which suggests that the galactic halo is much more massive than previously believed? [No reply.].

AN ABUNDANCE ANALYSIS OF FEHRENBACH'S STAR (HD 116745) IN OMEGA CENTAURI

R. J. DICKENS and A. L. T. POWELL*

Royal Greenwich Observatory, England

Abstract. HD 116745 is an eleventh magnitude early F-type giant occurring in the field of the globular cluster ω Cen. Membership in the cluster appears virtually certain since the star's radial velocity differs insignificantly from the mean cluster velocity of $+238$ km s^{-1}. HD 116745 lies nearly 4 mag. above the horizontal branch in the HR diagram and is presumably in a rapid, advanced stage of evolution. Spectrograms at 22.5 Å mm^{-1} and 155 Å mm^{-1} have been obtained by one of us (R.J.D.) at the Radcliffe Observatory in South Africa using a McGee spectracon image tube. The lower dispersion spectra have been used to measure Hγ profiles and the Balmer discontinuity, and one high dispersion film has provided the basis for a differential curve-of-growth analysis with respect to the Sun in order to determine some heavy element abundances. Physical data derived for the star are $M_v = -3.32$, $\theta_e = 0.77 \pm 0.02$, $\log g = 1.0 \pm 0.3$ and $\mathfrak{M} = 0.37\ \mathfrak{M}_\odot \pm^{0.37}_{0.18}$. Results of the curve-of-growth analysis, in which a model atmosphere was used to determine the curve appropriate to the star, yield a logarithmic iron-to-hydrogen abundance ratio with respect to the Sun of $[\text{Fe/H}] = -1.2$ with a provisional estimated uncertainty of about ± 0.2. This result is based on 28 lines of Fe I. Ti may be marginally overabundant, with a value of $[\text{Ti/Fe}] = +0.4$, based on 7 lines of Ti II. Other elements and ionization states identified are Mg I, Al I, Si I, Ca I, Sc II, V I, V II, Mn I, Sr II, Zr II (Eu II).

The small mass and possibly enhanced [Fe/H] above the mean for cluster giants (~ -1.7) are not unexpected on the basis of some current models of stars in late evolutionary stages in which most of the envelope mass might be lost, and an enhanced [Fe/H] could result from convective mixing from a region in which hydrogen burning is taking place. Such mixing might conceivably occur when the star is undergoing thermal pulses (driven by the He shell source instability), at times when the H-burning shell source periodically disappears (see, for example, the paper by Mengel at this Colloquium). However the present errors in the abundance and mass determinations are not yet small enough to rule out the possibility that HD 116745 has a mass typical of a horizontal branch star and a composition representative of an upper limit of a small range in abundance amongst the giant stars, which are known to show considerable intrinsic spread in colour in the HR diagram.

DISCUSSION

Bolton: Have you checked your equivalent width scale by taking spectra of the sun or some other star with the same equipment?

* Present address: Radcliffe Observatory, Pretoria, South Africa.

the 41-day cycle were hardly perceptible in 1963 and 1967, while they were quite strong in other years. We supposed at that time that these amplitude variations were irregular. To our surprise, in the fall of 1970 and early 1971 the amplitudes were again very small. In this way a 4-yr cycle became apparent. To study the transition from the old cycle to the expected new one we observed the ascending branch and maximum of RR Lyrae as many times as possible. The results for the visual maxima and phase-shifts of the median brightness are shown in Figure 1.

At the end of the old 4-yr cycle the phase variations during the 41-day cycle died down almost completely, and the new cycle then started with a rapid increase of the

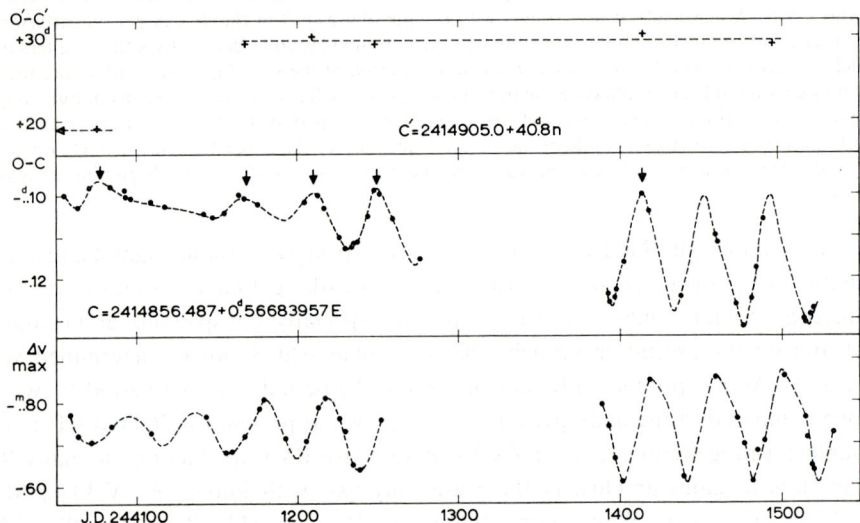

Fig. 1. The behavior of the 41-day cycle of RR Lyrae in 1971 and 1972, showing the development of a new 4-yr cycle. *Below*: The variation of the visual maxima of the $0^d.57$ period. *Middle*: The phase-variation of the median brightness ($V = 7.68$) on the ascending branch of the visual $0^d.57$ light-curve. The arrows indicate the most positive shifted ascending branches. *Above*: The $O' - C'$ values of the most positive shifted ascending branches calculated with $C' = $ JD 2414905.0 + $40^d.8\ n$. The new 4-yr cycle begins with a phase discontinuity of 10 days, after which $O' - C'$ remains constant.

amplitude. The amplitude of the maximum-variations was only $0^m.07$ at the end of the old cycle, and then very rapidly became as large as $0^m.16$ during 1971 and now is $0^m.27$. Most interesting is the phase-shift in the 41-day period. $O' - C'$ was $+19^d$ throughout the old cycle (1967–1971), while for the new cycle it is $+29^d$, i.e. the beginning of the new cycle was accompanied by a phase shift of 10 days, about a quarter of the 41-day period.

We have performed a similar analysis of the older data. The Konkoly Observatory has about 20 000 unpublished photographic observations from 1943 to 1949 and about 30 000 photoelectric observations between 1950–1953 and 1956–1972. Supplemented by other photoelectric observations in 1947 (Walraven, 1949), 1953 (Hardie, 1955),

1955 (Broglia and Masani, 1957), 1958–1959 (Cocito and Masani, 1960), 1961–1962 (Onderlička and Vetešnik, 1968), 1962–1964 (Preston *et al.*, 1965) and by the older visual and photographic material discussed by one of us (Detre, 1943) we were able to follow the 4-yr cycle back to 1935. The complete discussion of the entire material will be published elsewhere; here we show in Figure 2 only the amplitudes of the variation of the visual maximum and of the phase variation of the median brightness on the ascending branch of the $0^d\!.57$ light-curve during the 41-day period for the years 1943–1972, together with the phase-variations connected with the 41-day period.

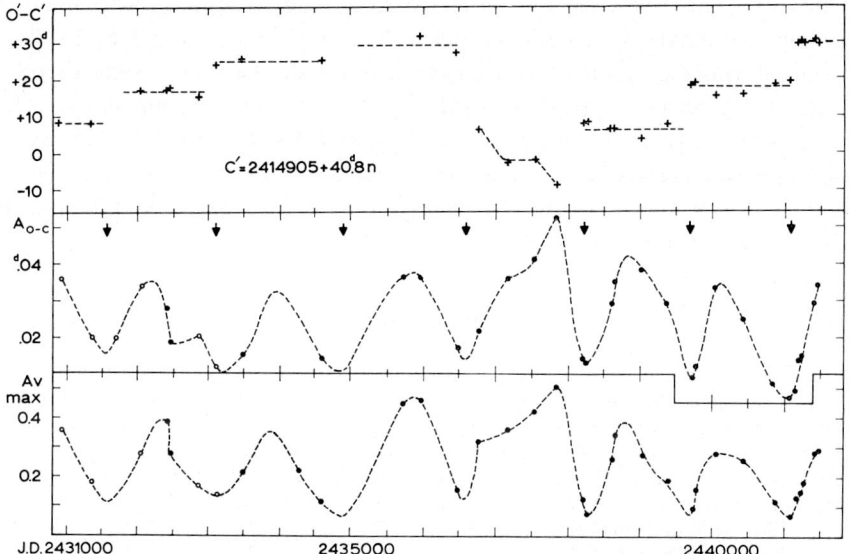

Fig. 2. *Below*: The amplitude of the variation of the visual maximum during the 41-day cycle vs JD. The 4-yr cycle is clearly shown. Open circles for photographic observations, approximately reduced to V magnitudes, filled circles for photoelectric V observations. *Middle*: The amplitude of the phase variation of the median brightness ($V = 7.68$) on the ascending branch of the $0^d\!.57$ light-curve during the 41-day cycle vs JD. The arrows indicate the beginnings of the different 4-yr cycles. *Above*: The $O' - C'$ values of the most positive shifted ascending branches calculated with $C' =$ = JD2414905.0 + $40^d\!.8n$. At the beginning of each 4-yr cycle a considerable phase discontinuity occurs, afterwards the $O' - C'$ values generally remain constant during one 4-yr cycle.

The amplitude variations are clearly cyclic, with cycle-lengths varying between 3.8 and 4.8 yr and having a mean value of 4.28 yr. The longest cycles have the largest amplitudes.

The most interesting feature of the 4-yr cycle is the considerable phase discontinuity in the 41-day period at the beginning of each 4-yr cycle. After this discontinuity $O' - C'$ remains constant during one and the same cycle; only the strongest cycle seems to be an exception.

Our results fit in well with Julia Balázs' interpretation of the 41-day cycle proposed 13 yr ago (Balázs, 1959; Balázs-Detre, 1962; Preston, 1964). The large observed phase shifts in the 41-day period are hardly understandable unless the 41-day period is equal to the rotation period of the star. Magnetic fields concentrated in a limited longitude zone around the equator develop and decay over a 4-yr interval causing 41-day variations in the light amplitude of the $0\overset{d}{.}57$ pulsation period and in the strength of the magnetic field as aspect effects of the 41-day rotation period of the star. The following 4-yr cycle starts with increasing magnetic fields in a different longitude, and this longitude difference between the new and old cycles gives rise to the observed phase shifts in the 41-day period. Curiously, the phase shift of 10 days, which has been observed several times corresponds to 90° in longitude.

In this way, the 4-yr cycle can be interpreted as the magnetic cycle of the star. Regarding the magnetic fields, Babcock's measures (Babcock, 1955, 1958) were performed during the maximum of a very high 4-yr cycle. They show some correlation with the 41-day period (Detre, 1961), the maximum positive and maximum negative fields being associated with maximum and minimum light amplitudes respectively. Preston's measures (Preston, 1967) in 1963 coincide with the minimum of a 4-yr cycle, those in 1964 with the beginning of a weak 4-yr cycle. That might be the explanation of why he has not once found a measurable field.

References

Babcock, H. W.: 1955, *Publ. Astron. Soc. Pacific* **67**, 70.
Babcock, H. W.: 1958, *Astrophys. J. Suppl* **3**, 141.
Balázs, J.: 1959, *Kleine Veröff. Remeis-Sternw.* No.27, p26.
Balázs-Detre, J. and Detre, L.: 1962, *Kleine Veröff. Remeis-Sternw.* No. 34.
Broglia, P. and Masani, A.: 1957, *Contr. Oss. Astr. Milano-Merate*, Nuova Serie, No.105.
Cocito, G. and Masani, A.: 1960, *Contr. Oss. Astr. Torino*, Nuova Serie, No. 27.
Detre, L.: 1943, *Mitt. Sternw. Ungar. Akad. Wiss.* No.17.
Detre, L.: 1961, *IAU Trans.* **XIB**, 295.
Detre, L.: 1969, *Non-Periodic Phenomena in Variable Stars*, Academic Press, Budapest.
Detre, L.: 1970, *Ann. Univ.-Sternw. Wien* **29**, No. 2, 79.
Hardie, R. H.: 1955, *Astrophys. J.* **122**, 256.
Onderlička, B. and Vetšnik, M.: 1968, *Astron. Inst. Univ. Brno (Czechoslovakia)*, Publ. No.8.
Preston, G. W.: 1964, *Ann. Rev. Astron. Astrophys.* **2**, 46.
Preston, G. W., Smak, J., and Paczyński, B.: 1965, *Astrophys. J. Suppl.* **12**, 99.
Preston, G. W.: 1967, in R. C. Cameron (ed.), *The Magnetic and Related Stars*, Mono Book Corp., Baltimore, p. 26.
Szeidl, B.: 1965 *Mitt. Sternw. Ungar. Akad. Wiss.* No. 58, p74.
Wachmann, A. A.: 1968, *Astr. Abh. Hamburg Sternw. Bergedorf* **8**, No. 114.
Walraven, Th.: 1949, *Bull. Astron. Inst. Neth.* **11**, 17.

VARIABLE STARS IN DWARF SPHEROIDAL GALAXIES

STEVEN VAN AGT

University of Nijmegen, Nijmegen, The Netherlands

1. Introduction

Interest in dwarf spheroidal galaxies is motivated by a number of reasons; an important one on the occasion of this colloquium is the abundance of variable stars. The theory of stellar evolution and stellar pulsations is now able to predict from theoretical considerations characteristic properties of variable stars in the colour-magnitude diagram (Iben, 1971). By observing the variable stars in the field, and in as wide a selection of objects as possible, more insight can be obtained into the history of the oldest members of our Galaxy (the globular clusters) and of the dwarf spheroidal galaxies in the Local Group. It is worthwhile to explore the spheroidal galaxies as observational tests for the theoretical predictions of conditions in space away from our Galaxy. The numbers of variable stars in the dwarf spheroidal galaxies are such that we may expect well-defined relations to emerge once reliable magnitude sequences have been set up, the variable stars found, and their periods determined. Six dwarf spheroidal galaxies are presently known in the Local Group within a distance of 250 kpc. In Table I, which lists members of the Local Group, they are at the low-luminosity end of the sequence of elliptical galaxies (van den Bergh, 1968).

How complete this number of spheroidals in the Local Group may be is uncertain. The discoveries are strongly affected by selection.

TABLE I
Local group members

Name	Type	M_v
M31 = NGC 224	Sb I–II	−21.1
Galaxy	Sb or Sc	−20
M33 = NGC 598	Sc II–III	−18.9
LMC	Ir or SBc III–IV	−18.5
SMC	Ir IV or Ir IV–V	−16.8
NGC 205	E6p	−16.4
M32 = NGC 221	E2	−16.4
NGC 6822	Ir IV–V	−15.7
NGC 185	dE0	−15.2
NGC 147	dE4	−14.9
IC 1613	Ir V	−14.8
Fornax	Spheroidal	−13.6
Sculptor	Spheroidal	−11.7
Leo I	Spheroidal	−11.0
Leo II	Spheroidal	− 9.4
Ursa Minor	Spheroidal	− 8.8
Draco	Spheroidal	− 8.6

The Sculptor dwarf galaxy was discovered quite accidentally by Shapley (1938a, b), and the Fornax system in a search of the southern polar-cap region on existing plates at Harvard Observatory. However the limiting magnitudes of these plates is only a little fainter than 18th magnitude and the plate scale is large. The Sculptor and Fornax dwarf galaxies are the only ones discovered in the southern sky. The northern sky has been searched systematically by Wilson (1955) on the plates of the Palomar Sky Survey.

The lack of a survey of the southern sky comparable in limiting magnitude and angular resolution to the Palomar Sky Survey greatly contributes to the uncertainties in the completeness. To find the extended Ursa Minor system even on the Sky Survey plates is difficult at galactic latitude $b = +45°$ due to the low stellar surface density in this galaxy. A ten times higher surface density in the field closer to the galactic plane would allow this spheroidal to completely escape detection.

The apparent lack of spheroidal galaxies at galactic latitudes smaller than $b = 35°$ is better explained as observational selection near our Galaxy than as another example of the apparent preference of dwarf galaxies to avoid low local galactic latitudes near large spirals (Holmberg, 1969). The recognition on the photographic emulsion of a dwarf spheroidal galaxy at large distances from the Galaxy is difficult (Reaves, 1956). Intrinsically faint galaxies have their light less concentrated towards the center and are therefore less easily recorded. The use of newly developed emulsions can prove rewarding, as is indicated by the recent discovery of three new dwarf spheroidal galaxies near M31 by van den Bergh (1972). The Kodak IIIaJ emulsion on which these dwarf spheroidals were discovered has an extremely high linear resolution and is especially suitable for recording faint magnitudes (Marchant and Millikan, 1965).

From dynamical considerations Hodge and Michie (1969) find it unlikely that the dwarf spheroidal galaxies ever were very close to the Galaxy, since close encounters would have completely disrupted the spheroidal, which is unimpressive in terms of mass: 10^5 to 5×10^6 $\mathfrak{M}_\odot$.

Estimates of the total number of dwarf spheroidal galaxies in the Local Group vary from 8 (Wilson, 1955) to more than 192 (van den Bergh, 1968), if one wishes to assume that spheroidal galaxies are homogeneously distributed out to 400 kpc in the Local Group.

Table II gives the positions in galactic coordinates of the nearby spheroidal galaxies.

TABLE II

Coordinates of nearby spheroidal galaxies

Name	l	b
Fornax	237°	−66°
Sculptor	286	−83
Leo I	226	+49
Leo II	219	+67
Ursa Minor	103	+45
Draco	86	+35

2. Variable Stars

For a full survey of the distribution of variable stars in extended dwarf spheroidal galaxies, wide-field telescopes can be used efficiently to avoid the numerous exposures off-set from the center of the galaxy necessary if telescopes of small angular field are used. The brightest stars in the nearest of the intrinsically faint dwarf galaxies reach approximately $B=17$ mag. The surface density of the stars down to the luminosity of the horizontal branch remains generally small, and for the central regions is 10 to 100 times higher than in the surrounding field. Telescopes with moderate angular resolution resolve the central regions well enough to allow inspection of the individual stars, and thus permit the search for and photometry of variable stars.

The first detailed investigation of individual stars in dwarf spheroidal galaxies was made by Baade and Hubble (1939) in the Sculptor galaxy. The authors reported 40 variable stars in the central region, but periods could not be derived from the limited number of 100-in. Mount Wilson plates. The range in stellar brightness and the distribution of the observations in time led Baade to the conclusion that the new variable stars were most likely RR Lyrae, with the exception of two variable stars with Cepheid characteristics.

Further investigation of the Sculptor and Fornax systems was prevented at that time by the lack of instrumental facilities in the south. In 1948 the 74-in. Radcliffe telescope became operational and new observations of the Sculptor galaxy were obtained by Thackeray. He discovered (Thackeray, 1950a) more than 230 variable stars in Sculptor. His material is well suited for deriving periods and is presently at the Dept. of Astronomy of the University of Nijmegen, the Netherlands, for further reduction.

After the discovery of the four spheroidal dwarf galaxies in the northern sky by Wilson (1955), Baade obtained large numbers of 200-in. plates of the dwarf spheroidal galaxies in Draco, Leo, Ursa Minor, and also of NGC 147 and NGC 185. His observations aimed at finding the variable stars now known to be present, and probably in large numbers, in the central regions of these objects, and at deriving their periods and magnitudes. For the Draco system the reductions are extensively discussed by Baade and Swope (1961). The Leo II dwarf galaxy is discussed in a progress report by Swope (1967) and the results for the Ursa Minor system by van Agt (1967).

New observations of the Sculptor system have been obtained recently by observers from the David Dunlap Observatory. These observations were made with the 24/36-in. Michigan Curtiss Schmidt telescope at Cerro Tololo, Chile. The $5° \times 5°$ field of the Curtiss Schmidt telescope, which has a plate scale of $97''.2$ mm^{-1}, accommodates the dimensions of the Sculptor system well.

Intercomparison of ten plate pairs of the Sculptor system yielded a rich harvest of over 500 variable stars. The total number of variable stars in this spheroidal galaxy, including those discovered earlier by Baade and Hubble, Thackeray and collaborators, Helen Hogg, who also blinked a plate pair, and the author is now over 600. The expected total number of variable stars is well over 850 as estimated from

TABLE V

Variable stars in some dwarf spheroidal galaxies

Name	Area searched in %	Total discoveries	RR Lyrae ab	RR Lyrae c	Cepheids $P<1^d$	Cepheids $P>1^d$	LPV	Irr red	Irr blue	Total measured
Fornax	100	–	–	–	–	–	–	–	–	–
Sculptor	100	603	54	9	2	2	1	1	–	68
Draco	75	261	126	7	2	2	–	–	–	138
Ursa Minor	26	92	21	9+4:	2	1	V 80	–	–	38
Leo I	–	–	–	–	–	–	–	–	–	–
Leo II	85	196	64	6	2	3	–	6	6	87

The reduction of the Fornax plates by Demers (1972) are not yet complete.

In the Sculptor dwarf spheroidal galaxy periods have been determined for a limited number of variable stars by Thackeray (1950b) and by van Agt (in preparation). Of the 63 variable stars with known periods 16% are c-type variables. This preliminary result indicates that in the Sculptor system c-type variables are probably not as scarce as in the Draco system.

In the field of Sculptor there are four short period variables with periods near one day. Included among these four is one variable with a period near $0\overset{d}{.}5$ and approximately 0.5 mag. brighter than the RR Lyrae stars in this dwarf galaxy.

Variable stars with periods of less than one day to over one day, but not exceeding $3\overset{d}{.}0$, and brighter than the other RR Lyrae stars in the dwarf spheroidal, are also in the Ursa Minor system (van Agt, 1967) and the Leo II system (Swope, 1967) and were discovered recently by Hodge in Leo I.

At a distance of 14' north of the centre of the Sculptor dwarf galaxy a long period variable star has been found on the Curtiss Schmidt plates. From the 74-in. plates we find that the time of rise to maximum and of decline to minimum is of the order of 120 days. A definite period has not been derived but observations over a long time interval indicate that a period near 150 days is possible.

The amplitude of the long period variable derived from plates with a IIIaJ emulsion exposed through a GG13 filter is 2.5 mag. The B_j magnitude is related to the BV system through the equation $B_j = B - 0.10(B-V)$. The B_j luminosity at maximum light is comparable to the brightest giant branch stars. The colour-magnitude diagram as derived by Hodge (1965) has a giant branch extending to large $(B-V)$ values. At least one of the very red stars at the tip of the giant branch is a variable star with small amplitude and probably irregular period.

The Ursa Minor system is the least rich in variable stars and also appears on the sky as the least well-defined. The c-type variables are relatively frequent in this Galaxy, forming 35% of the RR Lyrae population in the central region.

For one variable in the Ursa Minor system, V 80, the period is uncertain and classifying this variable as long period might be erroneous. The mean luminosity in B is about 0.5 mag. brighter than the RR Lyrae stars. A period close to one day is not excluded.

The Ursa Minor system contains among the bright short period variable stars two variables with periods well below one day and with a luminosity about 0.5 mag. brighter than the RR Lyrae stars.

The Draco system contains more than 260 variable stars (Swope, 1961) within a radius of 24' from the centre of the system, which has (Hodge, 1964) a limiting radius of 26'. The discoveries in the Draco system are complete within $6'.3$ from the centre and down to amplitudes in B of 0.4 mag. Only the stars in the central region have been measured. In the Draco system no long period or irregular variable stars have been discovered, although the distribution of the plates in time certainly would allow this.

The giant branch of the Draco system does not reach the very red colours observed in the colour-magnitude diagram of the Sculptor system.

At $B-V=1.4$ the giant branch is approximately 3.0 mag. brighter than the RR Lyrae stars. The giant branch is possibly double, but the number of stars in this region is small. From the luminosity of the giant branch it is expected that the stars in the Draco system are very metal poor. The very strong concentration of stars on the horizontal branch towards the red does not contradict this result since the distribution of the stars over the horizontal branch is not uniquely related to metallicity (Iben, 1971).

Variable stars brighter than the RR Lyrae stars have also been discovered in the Draco system.

The variable V 203, which has a small amplitude and is almost constant in luminosity during one observing season, falls above the horizontal branch in the colour-

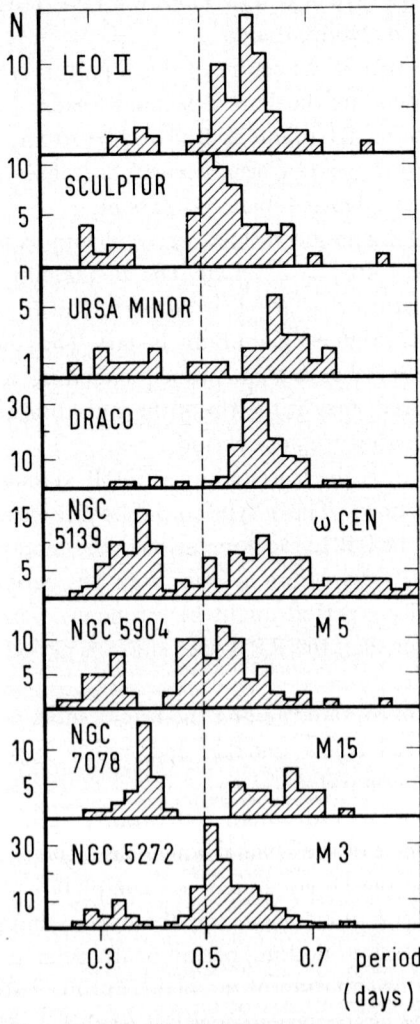

Fig. 1. Period-frequency diagrams for RR Lyrae stars in various dwarf spheroidals and globular clusters.

magnitude diagram; about where the main-sequence would have been if it continued upwards. It is not known if the star is irregular or eclipsing, or even if it belongs to the Draco system.

In the Leo II system, which has been investigated by Henrietta Swope (Swope, 1967), the total number of variable stars expected to be present was estimated by Baade at close to 250. 196 were discovered, and 87 of these have been measured. The percentage of c-type variables in this sample is only 9 percent. The colour-magnitude diagram, which is not calibrated in colour, shows a strong concentration of stars towards the red of the horizontal branch.

In the Leo II system the short period Cepheids with period less than one day are also brighter than the RR Lyrae stars in the same spheroidal.

For the Draco, Ursa Minor, Sculptor and the Leo II systems the period-frequency diagrams are brought together in Figure 1. In the same figure are also shown the period-frequency diagrams for the Galactic globular clusters NGC 5139, NGC 5272, NGC 7078 and NGC 5904. The data for the Leo II system are taken from unpublished data by Henrietta Swope.

Selection effects are at work in the diagram for the Sculptor system, where periods have been determined for a relatively small selection of variable stars. There is a possibility that those with larger amplitudes have been chosen preferentially. In the Ursa Minor system the number of variable stars in the central region is small. The period-frequency diagram for Sculptor is very similar to that for NGC 5272. In both diagrams the distribution is smooth and does not show double maxima. The distribution of the periods in the Draco system is comparable, but is shifted towards longer periods.

The distribution of the periods in the Draco system and the Leo II dwarf galaxy strongly disagrees with the concept of the two period groups for the Galactic globular clusters (van Agt and Oosterhoff, 1959). For Galactic globular clusters with sufficient numbers of variable stars, the mean period for the type-a variables in each cluster is for group I (the long period group), $P = 0\overset{d}{.}647 \pm 0\overset{d}{.}015$, and for group II (the short short period group) $P = 0\overset{d}{.}549 \pm 0\overset{d}{.}010$. The number of variable stars involved in the mean period of the RR_a variables in the dwarf spheroidal galaxies is sufficiently large to make the deviation from the mean periods of the two globular cluster groups meaningful.

For the dwarf spheroidal galaxies the mean periods are as shown in Table VI:

TABLE VI

Mean periods of RR Lyrae stars in spheroidal galaxies

Name	P_{mean}	N_c/N_{total}
Sculptor	$0\overset{d}{.}565$	0.14:
Draco	0.611	0.04
Ursa Minor	0.636	0.35
Leo II	0.592	0.09

The mean period of the RR_a variable stars in dwarf spheroidal galaxies clearly does not follow the pattern of the two period groups for Galactic globular clusters but has values inbetween those.

The period-amplitude relations for the dwarf galaxies are given in Figure 2, together with the period-amplitude relations for NGC 5139 and NGC 5272. The general pattern of decreasing amplitude with increasing period as observed for Galactic RR Lyrae stars is retained. Although one should be aware of the serious errors which

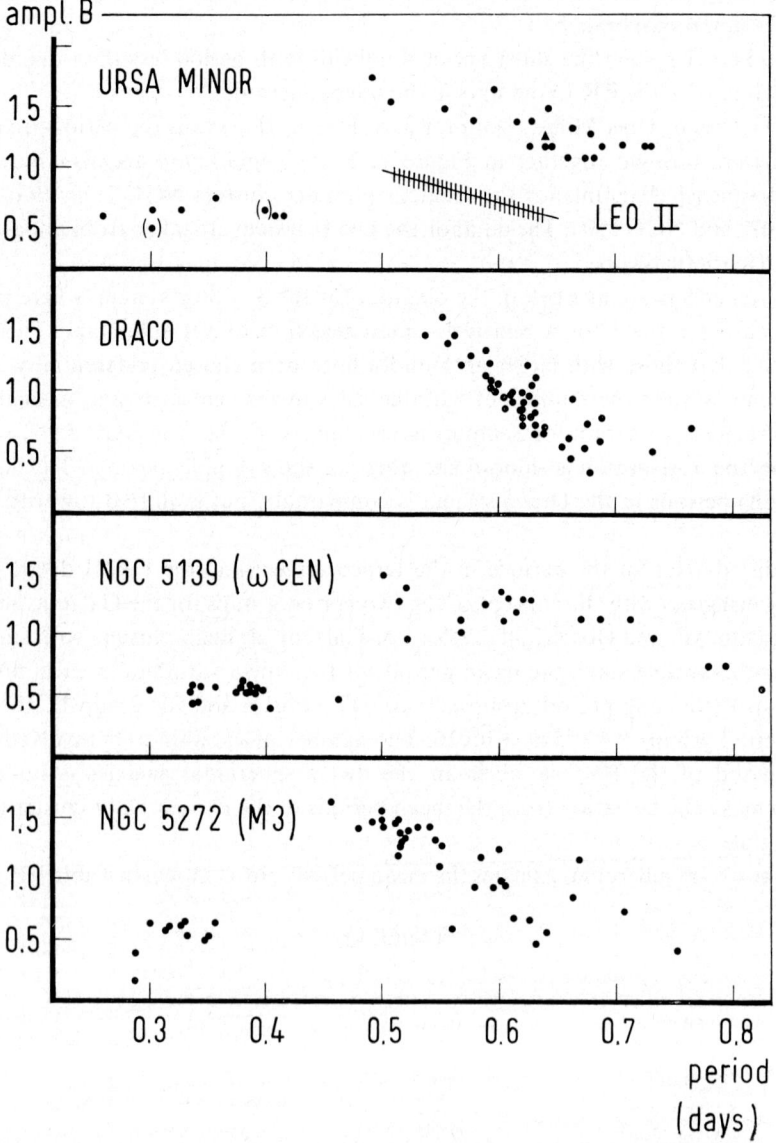

Fig. 2. Period-amplitude diagrams for RR Lyrae stars in various dwarf spheroidals and globular clusters.

can be introduced by incorrect magnitude scales, it seems evident from the figure that the slope of the period-amplitude relation varies distinctly from cluster to cluster. The slopes for the RR Lyrae stars in NGC 5139 and Ursa Minor system are comparable and, although the amplitudes reached in the Leo II system are considerably smaller, the slope for this dwarf spheroidal does not deviate much from the former two. The maximum amplitude in the Ursa Minor system is approximately 0.5 mag. larger than observed for the Leo II galaxy.

The Draco system is exceptional in the sense that the largest amplitudes are reached at longer periods than in either NGC 5272 or Ursa Minor. The slope of the period-amplitude relation is steeper and the relation is shifted as a whole towards longer periods.

If the conditions in the outer regions of the variable stars on the horizontal branch determine the amplitude of the brightness variations at a certain period then certainly a common factor must be at work for the variable stars in each cluster.

For the Sculptor dwarf spheroidal galaxy the period-amplitude relation is not yet available. However preliminary results for a number of variable stars, which are calibrated with the photographically transferred scale in the Small Cloud globular cluster Kron 3 (Walker, 1970) and NGC 121 (Tifft, 1963), indicate that the variable stars with periods near 0.5 day have amplitudes near 1.5 to 1.7 mag. These magnitudes are in the B_j system, derived from plates with IIIaJ emulsion exposed behind a GG13 filter.

The period-luminosity relation for the variable stars in the Draco dwarf galaxy is given by Swope (1961), and for the Ursa Minor system by van Agt (1967). For the Leo II system the magnitude scale is still provisional. The average apparent brightness of the RR Lyrae stars in five dwarf spheroidals is given below (Table VII).

TABLE VII

Average apparent magnitudes of RR Lyrae stars

Name	B_{RR}
Fornax	21.50:
Ursa Minor	20.32
Draco	20.48
Leo I	–
Leo II	21.90
Sculptor	20.60:

Henrietta Swope remarks in her discussion of the period-luminosity relation of the Draco system that there might exist a continuous relation between the stars with period longer than 0^d7 and the short period Cepheids in the dwarf spheroidal galaxies. The short period Cepheids in the dwarf spheroidal galaxies all have periods less than 2^d7 (V 6 in the Ursa Minor dwarf galaxy), and are systematically brighter by about one magnitude relative to the Cepheids of the same period in globular clusters.

between 17.9 and 18.8 in B and between 18.25 and 17.65 in V. The star is located at a distance of 8' from the center of NGC 121, for which the radius is somewhat smaller. The star might be a member of the Small Magellanic Cloud field.

Graham: It is most likely that immediate progress on obtaining good faint magnitude sequences near dwarf systems can be made using the Racine prism technique.

van Agt: The Racine wedge is extremely promising and we hope to obtain observations using this technique soon.

Schwarzschild: What is the frequency of occurrence of long period variables and the height of the giant branch?

van Agt: The only dwarf spheroidal which is known to contain a long period variable is the Sculptor system. The calibration of the giant branch points to possibly brighter stars at the tip of the branch relative to the RR Lyrae stars than generally found in metal poor globular clusters. However the calibration is still uncertain.

PART II

RR LYRAE VARIABLES
IN POPULATION II SYSTEMS

OBSERVATIONAL ASPECTS OF RR LYRAE VARIABLES IN GLOBULAR CLUSTERS

L. ROSINO

Astrophysical Observatory, Asiago, Italy

1. Introduction

RR Lyrae variables play a prominent role in many of the problems of globular clusters, and from several points of view. In the first place they can be considered as pretty good indicators of population and distance; although they do not form a completely homogeneous set of stars, the knowledge of their mean absolute magnitude gives a powerful means of establishing distances within and outside the Galaxy, and hence of determining the form and size of the Galaxy itself. Moreover, the number of RR Lyrae stars in clusters, the relative frequency of RR_c and RR_{ab} types, the length of the transition periods, the array of colors, when correctly interpreted, give important information on the degree of evolution, age and chemical composition of the clusters. Placed as they are in a peculiar region of the H−R diagram of Population II, the RR Lyr variables can be used as a good test of the theories of advanced evolution or the models of pulsating stars.

This explains why in the last ten years studies of the RR Lyr variables in globular clusters and nearby galaxies (Draco, Sculptor, Leo II), as well as of RR Lyr stars in the galactic bulge, have acquired so great an importance, and why they have been the object of observational research at many Observatories.

A survey of the present status of knowledge of RR Lyr variables in globular clusters from the observational point of view can be developed along the following lines:

(1) Frequency of RR Lyrae stars in globular clusters.
(2) Determination of magnitudes and colors.
(3) Periods and light curves; amplitudes.
(4) Secular and periodic variations of period and form of the light curves.
(5) Properties of RR Lyrae stars in different globular clusters.
(6) Absolute magnitude, composition, mass, age.

2. Frequency of RR Lyrae Stars in Globular Clusters

Of 122 known galactic globular clusters, about one-hundred have been searched for variables. However, not all of them have been thoroughly and extensively searched. An increase in the total number of variables may still be expected in some clusters, although it is very likely that the general situation will not be changed. A very important contribution in the survey for variables of southern globular clusters has been made by Fourcade and Laborde, of the Cordoba Observatory, who examined most of the clusters south of Declination −29° with the 60-in. telescope of Bosque Alegre.

Their Catalogue and Atlas (Fourcade and Laborde, 1966) are of extreme value for future investigations.

Sixteen globular clusters remain still unexplored (Table I). Some of them are Palomar clusters, very faint because of their large distance or strong interstellar absorption. The others are clusters in the direction of the galactic center, between 17 and 19 h in RA and $-8°$ to $-26°$ in Declination, strongly reddened. It is likely that most of these clusters, which belong to the nucleus-disk system and have an advanced spectral type, are poor in RR Lyrae variables. A survey should be opportune.

TABLE I

Globular clusters not examined for variable stars

Cluster	R.A.	Dec.	Class	$(m-M)_{app.}$
Pal 6	17^h40^m6	$-26°12'$	XI	21.5
Pal 8	18 38.5	$-19\ 52$	X	20.4
Pal 14	16 8.8	$+15\ 5$	–	–
Pal 15	16 57.6	$-\ 0\ 28$	–	–
Arp 2	19 25.6	$-30\ 27$	–	–
NGC 6316	17 13.4	$-28\ 5$	III	–
6325	17 15.0	$-23\ 42$	IV	–
6342	17 18.2	$-19\ 32$	IV	–
6355	17 20.9	$-26\ 19$	–	–
6440	17 45.9	$-20\ 21$	V	–
6517	17 59.1	$-\ 8\ 57$	IV	–
6544	18 4.3	$-25\ 1$	–	–
6638	18 27.9	$-25\ 32$	VI	–
6642	18 28.4	$-23\ 30$	–	–
6684	18 46.5	$-65\ 12$	–	–
6717	18 52.1	$-22\ 47$	VIII	17.1

Of one hundred clusters examined for variables only three contain more than 150 variables, mostly of RR Lyr-type. Two are the well known clusters NGC 5139 (ω Cen) with about 171 variables and NGC 5273 (M3) with 189 variables. The third, IC 4499, is a globular cluster studied by Fourcade and Laborde (1969) who found the system's exceptional richness in variables.

The sixteen globular clusters with more than 35 variables, the richest in the Galaxy, are listed in Table II. They contain nearly 1400 variables, of which more than 75% are RR Lyrae stars. Since the total number of variables in one hundred globular clusters is about 2060, the 16 clusters of Table II represent 70% of the total. It is therefore interesting to examine the properties of these clusters.

First they are all very rich in stellar population, as shown by their integrated absolute magnitudes (from -10 to -7.5) and as proved by counts of stars. However this is a necessary, but not a sufficient condition. We know in fact that extremely rich clusters, such as 47 Tuc and M13, are poor in variables.

Table II also indicates that, in clusters with many variables, most of the variables are RR Lyrae stars. Now, the occurrence of RR Lyrae stars depends on the density

TABLE II
Globular clusters with more than 35 variables

Cluster NGC (IC)	Var	RR	%RR	Sp.	Morgan class	Q	V	Oost. class	n_b/n
3201	82	72	88	–	–	−0.32	–	I	–
5139 (ω Cen)	171	138	81	F7	II	−0.39	2.9:	II	0.85
5272 (M3)	189	173	92	F7	II	−0.41	2.64	I	0.53
4499	170	–	–	–	–	–	–	–	–
5904 (M5)	97	92	95	F5	II	−0.39	2.58	I	0.72
6266 (M62)	89	74	83	F8	–	−0.31	–	I	–
6402 (M14)	77	63	82	F8	IV	−0.31	–	I	–
6715 (M54)	80	63	79	F7	III	−0.35	–	I	–
7006	72	67	93	F3–4	II	−0.40	2.6	I	–
7078 (M15)	102	74	74	F3	I	−0.44	3.1	II	0.73
5024 (M53)	47	33	70	F4	II	−0.37	3.1	II	0.85:
6121 (M4)	43	41	95	–	–	−0.31	2.5	I	0.40:
6934	51	44	86	F7	–	−0.35	--	–	–
6981 (M72)	39	39	100	G0	II	−0.32	2.6	I	0.60
2419	36	27	75	F5	–	−0.40	–	–	–
4590 (M68)	38	35	92	F2	–	−0.43	3.0:	II	–

and distribution of stars along the horizontal branch. The best condition seems to be reached when there is a moderate excess of blue over red components (ω Cen, M5) or an even distribution of stars (M3). There is only one cluster, NGC 7006, which is very rich in RR Lyrae variables and yet shows an excess of components to the red side of the gap. The peculiarities of the color-magnitude diagram of this remote cluster have been pointed out by Sandage and Wildey (1967).

Clusters rich in RR Lyrae variables have another property: most of them belong to the halo and have a relatively low metal abundance. With only one exception (M72, Sp. G3), their integrated spectral types are earlier than F8, and the corresponding Q is less than −0.31, the average value being −0.37. Most of these clusters belong to Morgan's class I–III with the sole exception of M4 (class IV).

It may be interesting to compare clusters rich in RR Lyrae stars with those in which no or few RR Lyrae stars have been found, notwithstanding the intrinsic richness of stars. Some of these variable poor clusters are listed in Table III. They can be divided into two classes: (a) clusters, like NGC 6254 (M10) or NGC 6205 (M13) which have

TABLE III
Rich clusters without or with few RR Lyr variables

Cluster NGC	Var	RR	%RR	Sp.	Morgan class	Q	V	Oost. class	n_b/n
104 (47 Tuc)	14	2?	–	G3	–	−0.26	2.15	–	0
6205 (M13)	15	2:	–	F5	III	−0.44	2.55	–	1
6218 (M12)	1	0	–	F7	–	−0.40	2.8	–	1
6254 (M10)	3	0	–	G0	IV	−0.41	2.85	–	0.9
6838 (M71)	4	0	–	G2	VI	−0.23	2.1	–	0
6637 (M69)	10	2?	–	G5	VII	−0.21	1.0	–	0

regular RR Lyrae variables from the simple formula $T_i = T_0 + iP$. It was pointed out a long time ago, however, that when early observations are discussed, the $O-C$, observed minus calculated phases, show in many cases systematic deviations which cannot be accounted for by observational errors. Since the RR Lyrae variables are evolving through the instability strip, and this means a change in the magnitude and color and therefore in the density and period, the occurrence of secular period changes was not surprising and gave the possibility of establishing the direction of evolution of RR Lyrae stars along the horizontal branch. Secular changes of period of the type $P = P_0 + \beta(t-T_0)$ give to the $O-C$ diagrams a parabolic form of the type: $O-C = \beta(t-T_0)^2/2P_0^2$. Parabolic curves for the $O-C$ were obtained by different observers, who derived values of β (variation of period in days per day) of the order of 10^{-10} ($0^d.03$ per 10^{-6} yr). Up to now, about eleven clusters rich in RR Lyrae stars have been examined for period variations. The results have not always been in agreement with the theories of horizontal branch evolution. In fact, it was noticed that in the same cluster some periods were increasing while others, without distinction of color, magnitude or period, were decreasing. For instance, examining 112 RR Lyr stars in Messier 3, Szeidl found that the periods of 22 variables were increasing with a rate of $0^d.18$ per 10^{-6} yr while 25 other variables showed periods decreasing at a rate of $-0^d.20$ per 10^{-6} yr.

In addition, some variables in M3 display $O-C$ curves which are by no means parabolic, but which indicate instead a periodic variation of the period (Figure 1.).

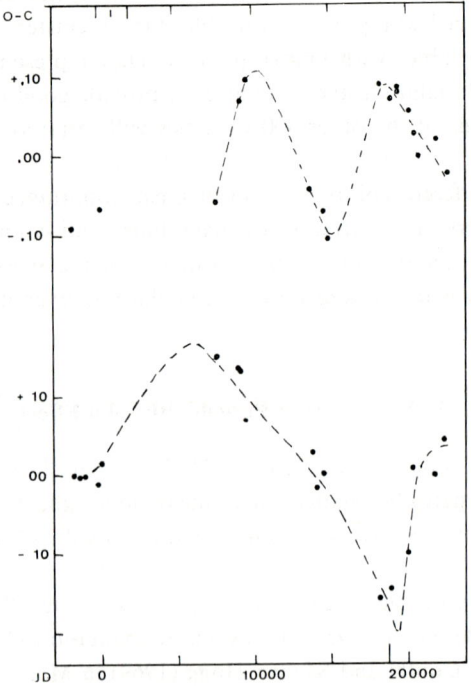

Fig. 1. Irregular $O-C$ curves observed by Szeidl (1965) in RR Lyrae variables of M3.

A somewhat different result was found by Belserene (1964) in an analysis of the RR Lyrae variables in Omega Centauri. She observed that 70% of the RR Lyrae variables show secular increases of period, while the other periods showed decreases, fluctuations, or remained constant. In M5, Coutts and Sawyer Hogg (1969) found 18 RR Lyrae variables with a constant period, 20 with increasing period ($0\overset{d}{.}05 \pm \pm 0\overset{d}{.}02\ 10^{-6}\ \text{yr}^{-1}$) and 12 with decreasing period. In M15 the stars with increasing periods outnumber those with decreasing period, but the tendency to increase is not as clearcut as in ω Cen.

Variations of periods in opposite directions are not easily accounted for, although Faulkner and Iben (1966) have found, by theoretical considerations, that stars *do* change direction of evolution in the horizontal branch. Moreover, by assuming that the stars do not spend more than 10^8 yr in the RR Lyrae strip, it is easy to show that the observed changes of period are sometimes at least one order of magnitude greater than expected. So, there is now a widely held opinion among observers that changes of period in RR Lyrae variables are not directly connected with the evolution of these stars, that they are not secular but periodic. The researches of Szeidl, Margoni and others have shown that periodic variations of the periods in the RR Lyrae variables of globular clusters are more frequent than previously believed. Abrupt changes of period or random changes may also occur so that the problem becomes even more complicated.

Periodic variations of period are closely connected with periodic changes in the form of the light curve. It is well known to variable star observers that the light curves of some RR Lyrae do not repeat exactly from cycle to cycle. While there are stars which undergo perfectly regular variations, others show changes in the phase of the rising branch, and in the shape and amplitude of the light curve (Blazhko effect). Szeidl (1965) has noticed, for instance, that of 112 variables in M3, at least 36 show the Blazhko effect. Stars with variable light curves are encountered more frequently in M3 among the RR Lyrae stars with periods between $0\overset{d}{.}47$ and $0\overset{d}{.}56$. Some years ago, Detre (1961) pointed out that there are two different types of Blazhko effect. In the first type the lower part of the rising branch is constant in time and phase, while the upper part changes from one cycle to another in phase and amplitude. In stars of this type phase variations are greater at maximum than at minimum. In the second type, on the other hand, phase variations are small near maximum, but pronounced in the lower part of the light curve. This type is more frequently encountered among variables having a long-term secondary period. Different cycles may differ in length and amplitude in the variables affected by the Blazhko effect. An example is shown by the RR Variable No. 30 in M53 studied by Margoni and also by Wachmann (1968). Periodic changes of the light curve and brightness at maximum occur in this variable with a period of about 37 days (Figure 2).

Work on the period variations and Blazhko effect in the RR Lyrae stars of globular clusters is just beginning. It should be extended and refined with photoelectric observations in different colors, for their obvious implications in pulsation theory and stellar models. Particular care should be taken in the search for incipient or final stages of

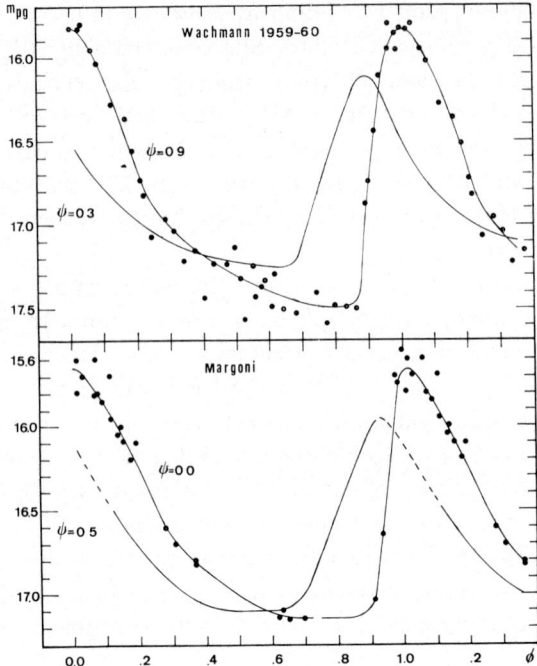

Fig. 2. Variations in the light curve and amplitude of variable No.30 in M15 due to the Blazhko effect, observed by Margoni and Wachmann (1968) at different phases of the 37-day period.

variability at the boundaries of the instability strip, and in the observation of stars near the transition from the fundamental to the first overtone mode.

6. General Properties of RR Lyrae Variables in Globular Clusters

It is well known that globular clusters may differ among themselves not only in age, concentration and richness of stars, but also in chemical composition, particularly for the relative abundance of hydrogen, helium and heavy elements. Theoretical work on unmixed horizontal branch models have shown that metal abundance may strongly affect the size and extension of the horizontal branch, and hence also the color and period of the RR variables. Also other parameters exert a strong influence on the horizontal branch. An observational approach to the theory should therefore follow these lines: (a) Establish properties of the RR Lyrae variables within rich clusters, as for instance ω Cen, M3, M5, M15. (b) Compare the RR Lyrae in clusters of different characteristics and see how their properties depend on the parameters of the cluster.

RR Lyrae stars in clusters are distributed in two classes: RR_c stars with periods less than $0^d.4$, small amplitudes, light curves of sinusoidal type, and RR_{ab} stars with periods longer than $0^d.4$, amplitudes up to $1^m.7$ and even more, asymmetric light curves with steep rise and slow decline. It is generally assumed that *ab* variables vibrate in the fundamental mode and c variables in the first overtone, the ratio of the fundamental to the first overtone period being about $\frac{4}{3}$. The two classes of stars appear neatly separated on a *period-amplitude* diagram. The RR_c stars maintain more or less the same amplitude (about $0^m\!.5$ in ω Cen) while the RR_{ab} stars show a decrease in amplitude with increasing period. RR_c variables have been further subdivided into two subtypes: those with periods from $0^d\!.20$ to $0^d\!.36$ and those with periods from $0^d\!.36$ to about $0^d\!.40$. In the *period-color* diagram the two subtypes occupy different positions, the shortest period group being to the left, near the blue edge of the strip. The RR_{ab} stars have also been classified into two subtypes: those with $P<0^d\!.6$ and $P>0^d\!.6$. The long period group has, at equal period, a larger amplitude and a bluer color than the short period group. The appearance of the diagrams suggests that a cluster may contain both groups of RR_{ab} variables, but in different proportions (Figure 3).

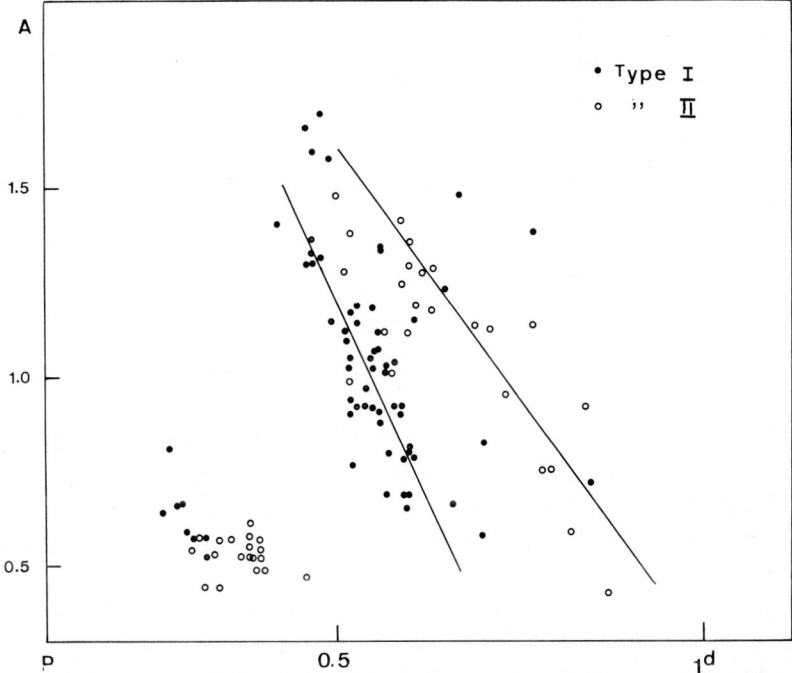

Fig. 3. Period-amplitude relation for R R I yr variables in globular clusters of Oosterhoff type I and II.

The *period-luminosity* relation of RR Lyrae stars is also very interesting. At first sight, all RR Lyrae variables in a given cluster have about the same luminosity. However, when precise measures of the mean magnitudes are made, it is apparent that both RR_c and RR_{ab} increase slightly in brightness with period (Figure 4). The positions of the two types in the color-magnitude diagram have already been discussed. The separation in color of the two types may or may not be sharp (Figure 5). In general, however, the color index increases with period, the RR_c stars being in the

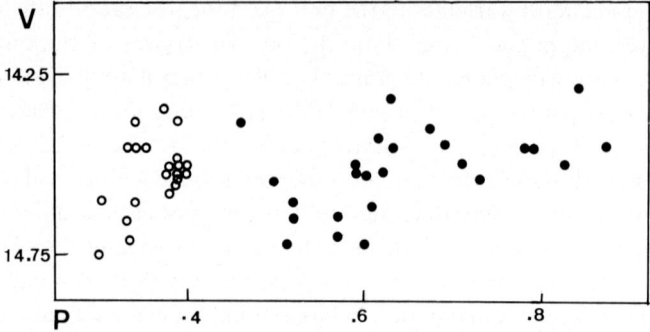

Fig. 4. Period-luminosity relation for RR Lyr variables of ω Cen (Dickens and Saunders, 1965).

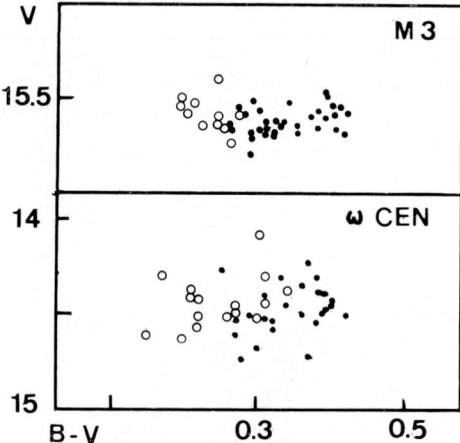

Fig. 5. Color-magnitude diagram for RR Lyr variables in M3 and ω Cen (Geyer and Szeidl, 1970).

average bluer than the *ab* stars. The period-amplitude, period-color and color-amplitude relations indicate some differences from cluster to cluster. When, however, the frequency distribution of the two types and the distribution of their periods are considered, these differences suddenly become very important. It is well known (Oosterhoff, 1939, 1944; Sawyer Hogg, 1944) that galactic globular clusters can be divided into two well-separated groups, according to the number N and mean period $\langle P \rangle$ of the RR_{ab} and RR_c types. In group I the mean period of RR_{ab} stars is nearly $0^d.53$ and that of RR_c stars is $0^d.32$ (Figure 6). The ratio $r = N_c/(N_c + N_{ab})$ is on the average 0.18. In group II the corresponding periods are respectively $0^d.65$ and $0^d.37$ and the ratio $r \sim 0.44$. In other words, in group I there is a predominance of RR_{ab} over RR_c stars and the mean periods are shorter than in group II, where RR_c stars are about as numerous as the RR_{ab} ones. Table IV illustrates the characteristics of the two groups according to the most recent data. Some extreme cases are presented by NGC 6981 with 27 RR_{ab} stars and only one RR_c star and in M68 with 21 RR_c stars and only 14 RR_{ab} types. Selection effects do not influence the relative abundance of

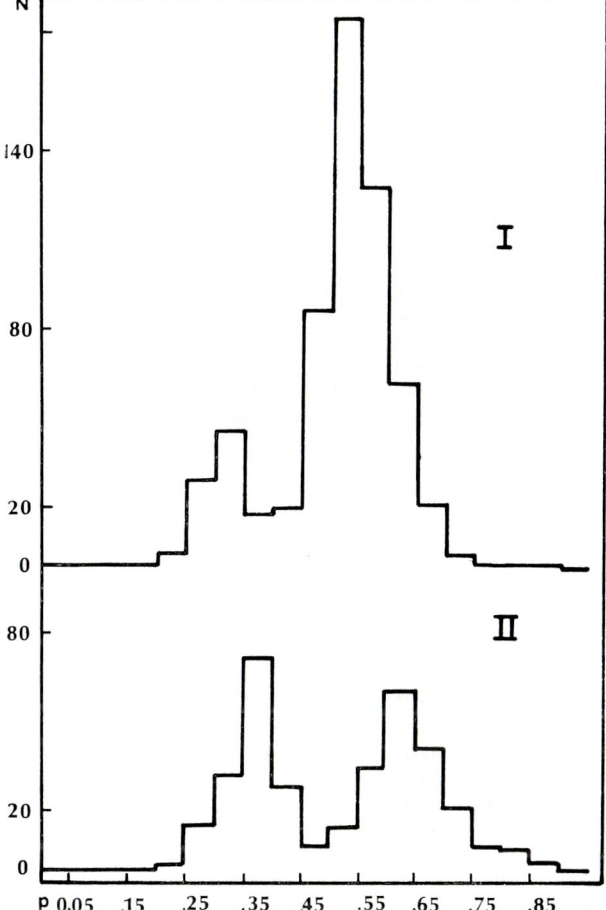

Fig. 6. Period distribution of RR Lyr variables in Oosterhoff groups I and II.

RR_c stars in the two groups, since the amplitude is about the same and the probability of discovery depends on the amplitude. A unique case is that of the loose cluster Pal 5 (not included in Table IV) which has five RR Lyrae variables, all of type c.

The classification of the globular clusters into two groups can be refined. Castellani *et al.* (1970) have proposed a multidimensional classification of the clusters into four classes, considering more strictly the ratio of type *ab* to *c*. Group I is formed by subclasses AI with a predominance of *ab*'s over *c*'s, and CI with about an equal number of *ab*'s and *c*'s (NGC 4147, 6362); group II is subdivided into CII with about the same number of *ab*'s and *c*'s, and AII with a predominance of *ab*'s over *c*'s (M9, M2). Besides the quantity S of Hartwick (1968) and the color index of the junction point $(B-V)_{0,g}$, they introduce as significant parameters the ratio N_c/N_{ab} and the absorption free parameter Q. This last parameter is shown in Table VI, together with the integrated spectral type of the cluster and the parameter ΔV, measured at $(B-V)_0 =$

7. Absolute Magnitudes, Masses and Radii of RR Lyrae Stars in Globular Clusters

For many years RR Lyrae variables have been considered as the best distance indicators within the Galaxy. It was assumed that all RR Lyrae variables, wherever they were, had the same absolute photographic magnitude, equal to zero. However, after the main sequences of some globular clusters were identified, it became apparent that this assumption was incorrect. The fitting of the main sequences, even when due corrections for interstellar absorption and blanketing were applied, gave a dispersion of the mean absolute magnitudes of the horizontal branches which was significant. Additionally, a difference in the absolute magnitudes of the RR Lyrae variables was also a prevision of the theories of stellar evolution, considering that the position of the horizontal branch depends on the age and composition of the cluster concerned.

An approach to the determination of the absolute magnitudes of RR Lyrae variables was made from various directions. Statistical parallaxes were determined for galactic RR Lyrae stars in moving groups by Eggen and Sandage (1959, 1962), who obtained an average value $\langle M_v \rangle = +0.6$. More recently Woolley and Savage (1971), from a discussion of the secular parallaxes of a number of galactic RR Lyrae stars, derived values of the visual absolute magnitude ranging from $+0.62$ for RR_c types with $P > 0^d.36$ to $+0.46$ for RR Lyrae ab types with $P > 0^d.44$.

In 1965, after a careful determination of reddening and color excess, Sandage (1969) arrived at the following values of the visual absolute magnitude of RR Lyrae stars in different clusters by a main sequence fitting procedure:

M92	+0.47
M15	+0.51
M13	−0.09
M3	+0.38
47 Tuc	+0.44.

These values were later improved by Sandage (1970) using field subdwarfs with known trigonometric parallaxes for calibration of the main sequences. The final result was $M_v = +0.6 \pm 0.2$ for the mean absolute magnitude of the RR Lyrae variables in M3 (type I), M15 and M92 (type II). The result, however, was somewhat disappointing, since the RR Lyrae stars in M3 were found to be about $0^m.3$ brighter than the RR Lyrae stars of M92 and M15, while the pulsation theory gives a difference of $0^m.3$, but in the opposite direction. So, the observations have come to exactly the opposite result than theory. The RR Lyrae variables in M13 were found to be much brighter ($M_v = +0.05$), than the others but this fact was easily explained by assuming that they might be associated with the asymptotic branch rather than the horizontal branch.

There are other methods, based on pulsation theory, for deriving the absolute magnitudes of RR Lyrae variables in globular clusters. Christy (1966) has found that the transition period from the fundamental mode to the first overtone is dependent on the luminosity according to the equation:

$$P_{tr} = 0.057(L/L_\odot)^{0.6},$$

which can be reduced to the following (Dickens, 1970):

$$M_v = -0.46 - 4.17 \log P_{tr}.$$

Using this formula, and with the values of the transition period reported in the last column of Table IV, Dickens derived from a grouping of the clusters in the Deutsch-Kinman A, B, C classes, the following mean values for the mean absolute magnitudes:

Very low metal abundance
(class C) $M_v = +0.51$
Low metal abundance
(class B) $+0.69$
Moderate metal abundance
(class A) $+0.96$

By grouping the clusters into the Oosterhoff classes I and II, the following values of the mean absolute visual magnitude have been found:

Group I $M_v = +0.926$
Group II $.559$

with a difference $\Delta M_v = +0.367$ between group I and group II. It should be kept in mind that these results assume the full validity of the pulsation theory. If the observational errors in the direct determination of the absolute magnitudes of the RR Lyrae stars could be reduced, or if the distances of the systems containing RR Lyrae could be determined by other methods there would be the possibility of obtaining more reliable values for the absolute magnitudes, and, at the same time, of obtaining further controls on the pulsation theory. At the present moment, the only conclusion which can be drawn from direct observations is that the mean visual absolute magnitude of RR Lyrae stars in globular clusters is about $M_v = +0.6 \pm 0.2$.

Another point concerns the dispersion of the magnitudes of the RR Lyrae stars within the same cluster, which can reach as much as $0^m.3$. Part of this dispersion is certainly due to observational errors or the Blazhko effect, but in part it is intrinsic. The dispersion may be ascribed to differing stages of evolution or different composition among the stars of the horizontal branch.

Radii and masses of galactic RR Lyrae stars have been derived recently by Woolley and Savage (1971) through an extension of the Baade-Wesselink method, taking into account the surface gravity and its variations. For RR_{ab} stars with $P > 0^d.44$, absolute magnitude $+0.40$, they have found $R \sim 5.5 R_\odot$, $\mathcal{M} \sim 0.5 \mathcal{M}_\odot$; for RR_c stars with $P > 0^d.36$, $M_v = +0.8$, $R \sim 4.5 R_\odot$, $\mathcal{M} \sim 0.6 \mathcal{M}_\odot$. From these parameters, values of the pulsation constant are found to be in fairly good agreement with those given by the pulsation theory.

Masses can also be derived from the pulsation theory, from knowledge of $(B-V)_{trans}$ which gives the T_e at the transition point, and from the high temperature boundary of the instability strip, given by the color index of the bluest c-type variables. This last parameter also gives the helium content. The values derived by Dickens for three globular clusters: NGC 6171, M3, ω Cen are: Masses from 0.48 to 0.43 solar masses; M_V from 0.57 to 1.10; helium content from 0.33 to 0.46. These values confirm those found by Sandage, who gives a helium content from 0.3 to 0.35 and masses of the order of 0.5. The ages of globular clusters, derived from the magnitude of the turn-off point are, according to Sandage, about 10×10^9 yr, in good agreement with cosmological theories.

References

Belserene, E. P.: 1952, *Astron. J.* **57**, 237.
Belserene, E. P.: 1964, *Astron. J.* **69**, 475.
Castellani, V., Giannone, P., and Renzini, A.: 1970, *Astrophys. Space Sci.* **9**, 418.
Castellani, V., Giannone, P., and Renzini, A.: 1972, Private communication.
Christy, R. F.: 1966, *Astrophys. J.* **144**, 108.
Coutts, C. M. and Sawyer Hogg, H. B.: 1969, *Publ. David Dunlap Obs.* **3**, 1.
Detre, L.: 1961, *IAU Reports on Astronomy* **XIB**, 295.
Dickens, R. J.: 1970, *Astrophys. J. Suppl.* **22**, 187.
Dickens, R. J. and Saunders, J.: 1965, *Roy. Obs. Bull.* No. 101.
Eggen, O. J. and Sandage, A.: 1959, *Monthly Notices Roy. Astron. Soc.* **119**, 255.
Eggen, O. J. and Sandage, A.: 1962, *Astrophys. J.* **136**, 735.
Faulkner, J. and Iben, I.: 1966, *Astrophys. J.* **144**, 995.
Fourcade, C. R. and Laborde, J. R.: 1966, *Atlas y Catalogo de Estrellas variables en cumulos globulares al sur de* $-29°$, Cordoba.
Fourcade, C. R. and Laborde, J.: 1969, *Mem. Soc. astr. Ital.* **40**, 1.
Geyer, E. H. and Szeidl, B.: 1970, *Astron. Astrophys.* **4**, 40.
Hartwick, F. D. A.: 1968, *Astrophys. J.* **154**, 475.
Kukarkin, B. V.: 1961, *Trans. IAU*, **XIB**, 300.
Margoni, R.: 1965, *Contr. Oss. astrofis. Univ. Padova* No. 170.
Margoni, R.: 1967, *Contr. Oss. astrofis. Univ. Padova* No. 198.
Martin, W. Ch.: 1938, *Ann. Sterrew. Leiden* **17**, II.
Oosterhoff, P. Th.: 1939, *Observatory* **62**, 104.
Oosterhoff, P. Th.: 1941, *Ann. Sterrew. Leiden* **17**, IV.
Oosterhoff, P. Th.: 1944, *Bull. Astron. Inst. Netherlands* **10**, 58.
Sandage, A.: 1958, in D. J. K. O'Connell (ed.), *Stellar Populations*, North Holland Pub. Co., Amsterdam, p. 41.
Sandage, A. and Wildey, R. L.: 1967, *Astrophys. J.* **150**, 469.
Sandage, A.: 1969, *Astrophys. J.* **157**, 515.
Sandage, A.: 1970, *Astrophys. J.* **162**, 841.
Sawyer Hogg, H. B.: 1944, *Commun. David Dunlap Obs.* No. 11.
Szeidl, B.: 1965, *Mitt. Sternw. Ungar. Akad. Wiss.* No. **58**, p 265.
Wachmann, A. A.: 1968, *Astr. Abh. Hamburg. Sternw. Bergedorf* **VIII**, 4.
Wesselink, A. J.: 1969, *Monthly Notices Roy. Astron. Soc.* **144**, 297.
Woolley, R. and Savage, A.: 1971, *Roy. Obs. Bull.* No. 170.

DISCUSSION

Schwarzschild: Is is not possible to discover very small amplitude variables by careful photographic photometry?

Rosino: Discovery of variables in globular clusters with an amplitude less than $0^{m}\!.2$ is very difficult when the usual techniques (blink, negative on positive, and so on) are employed. However, if a given star is, for some reason, suspected of variability and special and extensive observations of the particular star are made then even a variation less than $0^{m}\!.2$ can be detected.

Dickens: Typical standard errors (per plate) are about 0.03 mag. at best so that variables with amplitudes less than about 0.2 mag. would be difficult to find, except perhaps with rather specialized and extensive measurements.

Cox: Is the colour boundary extremely sharp between c type variables and non-variables?

Dickens: Only in M3 are the observations adequate to test this and indeed the separation between variables and non-variables is quite sharp, as shown many years ago by Roberts and Sandage.

Graham: Bright field horizontal branch stars can be picked out by Stromgren *uvby* photometry alone. Careful examination with photoelectric photometry of those stars whose colors indicate that they are near to the instability strip could possibly detect small amplitude RR Lyrae stars. Examination of one such star, HD 161817 by Graham and Zinter some years ago failed to reveal any variations as great as $0\overset{m}{.}1$.

Jones: I know of no small amplitude RR Lyrae in the field. The small amplitude variables have shorter periods and only a few have weak lines.

Cox: I believe that non-linear pulsation theory predicts a very sharp blue edge to the pulsation instability strip. Linear theory certainly does. In a given cluster, where stars exist to check this point, how sharp is this blue edge observed to be?

Rosino: I think it is of the order of one tenth of a magnitude, but it strongly depends on the observational technique.

NEW VARIABLE STARS IN THE GLOBULAR CLUSTER NGC 6401*

A. TERZAN and B. RUTILY

Lyons Observatory, France

Abstract. 115 new variable stars have been detected in a one square degree field centered on the globular cluster NGC 6401. Two of them (No. 157 and 164) seem to belong to the cluster.

1. Introduction

From its first observation in 1784 by Herschel (1786) till now, the globular cluster NGC 6401 ($\alpha = 17^h 35^m.53$; $\delta = -23°52'.89$; 1950.0) has not been systematically searched for variable stars, either in the cluster itself or in the surrounding field (Alter *et al.*, 1970; Hogg, 1963).

The importance of interstellar absorption in front of the cluster is perhaps the reason: 3.44 mag. if $A_v = 0.24 \operatorname{cosec} b$ (Arp, 1965), or 2.69 mag. with $D'_{0.9}$ (Kron and Mayall, 1960).

In 1968 and 1970 we made a large number of photographic observations at the newtonian focus of the 80-cm reflector ($F/6$) at Haute Provence Observatory. The plates are centered on the cluster and cover a field of about one square degree. The comparison between them with the blink microscope of the Lyons Observatory led us to discover 102 new variable stars, one (No. 41) of which seems to belong to the cluster (Terzan and Rutily, 1972). They are the first variables to be discovered in the field of NGC 6401 (Kukarkin, 1971).

Photographic observations (103 aE plates + Ilford 204 filters, $\lambda_{\text{eff}} \cong 6400$ Å) have been pursued at the newtonian foci of the 80-cm ($F/6$) and 193-cm ($F/5$) reflectors at Haute Provence Observatory in 1971 and 1972.

In this work, we give the latest results obtained from these new observations.

2. Observations and Measures

The many plates obtained since 1968 have been inter-compared with the blink microscope of the Lyons Observatory. The photometric measures are made with an iris-photometer (accuracy: ± 0.04 mag.).

The sequence of red magnitudes (m_r) used for this study is the one recently established around the cluster (Terzan and Rutily, 1972). 115 other new variable stars have been detected near NGC 6401. They add to the 102 variables found previously (Terzan and Rutily, 1972) and are numbered from 103 to 217 (Figures 1 and 2).

---

* Observations made at Haute Provence Observatory (C.N.R.S.).

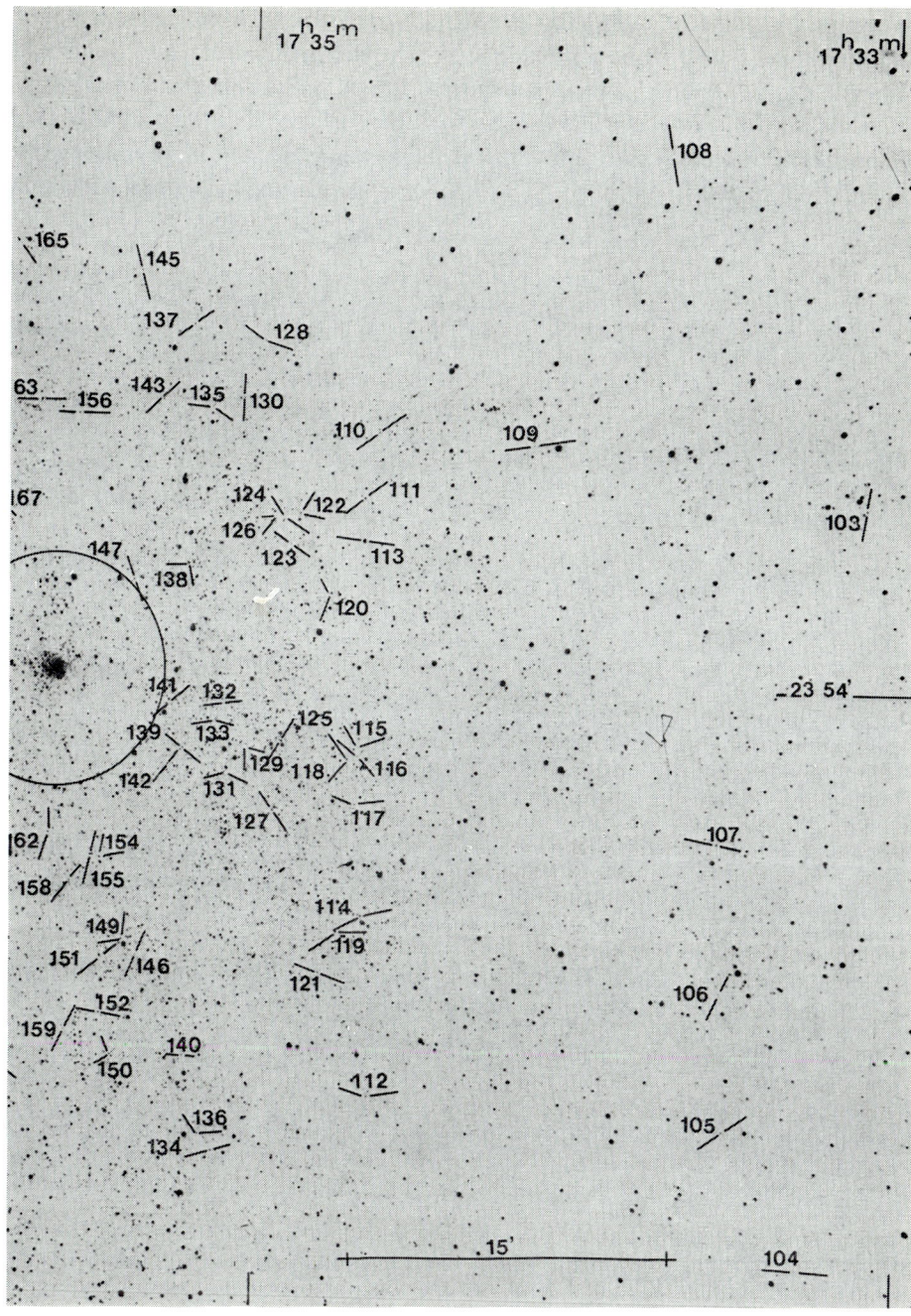

Fig. 1. New variable stars (103–159) detected near the globular cluster NGC 6401.

TABLE I
Positions and magnitudes of new variable stars

No.	α (1950.0)	δ	m_1	m_2
103	17h32m88	−23°45.′85	15.6	16.1
104	33.08	−24 18.81	15.8	>17
105	33.33	−24 12.84	15.9	16.4
106	33.34	−24 06.83	16.1	16.6
107	33.35	−24 00.37	15.2	15.7
108	33.54	−23 30.52	16.1	>17
109	33.98	−23 43.21	16.3	16.9
110	34.49	−23 42.71	15.9	16.8
111	17 34.51	−23 46.10	15.7	16.8
112	34.51	−24 11.30	15.7	17.0
113	34.54	−23 47.37	12.4	13.2
114	34.54	−24 03.51	15.8	16.6
115	34.55	−23 56.26	16.4	>17
116	34.55	−23 56.63	16.5	>17
117	34.56	−23 58.74	15.7	16.5
118	34.59	−23 56.77	15.4	16.0
119	34.62	−24 04.22	15.4	16.1
120	34.65	−23 49.65	16.2	16.8
121	17 34.68	−24 05.96	15.9	17.0
122	34.75	−23 46.16	15.7	16.3
123	34.78	−23 47.50	16.3	17.0
124	34.80	−23 46.33	16.2	>17
125	34.81	−23 55.96	15.9	16.5
126	34.83	−23 46.29	16.5	>17
127	34.83	−23 58.97	15.1	15.7
128	34.87	−23 38.75	16.2	16.7
129	34.93	−23 56.25	16.3	>17
130	34.94	−23 41.23	15.4	16.4
131	17 34.99	−23 57.33	15.8	16.3
132	35.00	−23 54.39	14.6	>17
133	35.02	−23 55.08	16.6	>17
134	35.03	−24 13.68	15.6	16.1
135	35.04	−23 41.64	15.9	16.8
136	35.07	−24 12.94	15.8	16.2
137	35.11	−23 38.06	15.9	>17
138	35.12	−23 48.37	15.0	15.6
139	35.13	−23 56.28	15.7	16.2
140	35.13	−24 09.54	16.2	17.0
141	17 35.18	−23 54.41	16.4	17.0
142	35.18	−23 57.00	15.6	16.1
143	35.22	−23 41.31	16.2	16.8
144	35.25	−23 52.74	15.2	15.6
145	35.28	−23 35.79	15.9	16.7
146	35.29	−24 05.24	16.3	17.0
147	35.30	−23 49.05	14.1	16.1
148	35.32	−23 49.55	16.0	16.5
149	35.33	−24 04.56	15.7	16.1
150	35.37	−24 09.50	15.9	>17

Table I (Continued)

No.	α (1950.0)	δ	m_r	
			m_1	m_2
151	17ʰ35ᵐ38	−24°05′.66	16.3	>17
152	35.40	−24 07.71	15.4	15.9
153	35.41	−23 50.26	15.8	16.8
154	35.41	−24 01.04	15.5	16.0
155	35.44	−24 00.92	15.6	16.1
156	35.48	−23 41.94	15.4	16.3
157	35.51	−23 52.82	15.9	16.5
158	35.51	−24 02.03	15.8	16.7
159	35.52	−24 08.45	16.0	>17
160	35.55	−23 54.95	15.5	15.9
161	17 35.58	−23 54.88	15.9	16.4
162	35.58	−23 59.96	15.3	16.1
163	35.61	−23 41.38	15.0	15.8
164	35.63	−23 52.93	15.2	15.9
165	35.68	−23 34.71	16.0	16.7
166	35.72	−23 51.82	16.5	>17
167	35.74	−23 45.56	15.7	>17
168	35.75	−24 09.75	15.4	16.2
169	35.77	−23 49.57	15.7	16.4
170	35.77	−23 52.47	14.2	14.6
171	17 35.81	−23 50.78	17.0	>17
172	35.83	−24 02.08	15.7	16.2
173	35.99	−23 56.22	15.5	16.2
174	36.00	−23 48.53	16.1	17.0
175	36.03	−23 43.06	16.0	17.1
176	36.10	−23 36.89	15.9	16.5
177	36.10	−23 58.03	15.3	16.0
178	36.11	−23 49.86	15.5	16.3
179	36.14	−23 48.51	15.4	16.1
180	36.17	−23 42.52	15.7	>17
181	17 36.17	−24 05.34	15.5	16.5
182	36.20	−23 49.62	16.2	>17
183	36.22	−23 31.02	15.9	16.5
184	36.25	−23 34.32	15.5	16.7
185	36.34	−23 39.86	15.7	>17
186	36.35	−24 05.69	15.8	16.8
187	36.36	−23 58.75	15.1	15.8
188	36.38	−23 55.90	15.2	>17
189	36.44	−24 07.38	15.8	>17
190	36.46	−24 01.81	14.9	16.0
191	17 36.49	−24 06.82	15.5	16.6
192	36.50	−24 08.64	14.7	15.3
193	36.53	−23 57.46	15.9	16.8
194	36.71	−24 09.12	15.1	15.6
195	36.83	−23 35.60	15.1	>17
196	36.88	−23 54.49	16.3	16.8
197	36.88	−23 55.11	16.1	>17
198	36.91	−23 45.37	15.7	16.1
199	36.99	−23 43.43	15.3	15.8
200	37.02	−23 55.56	16.0	16.7

Table I (Continued)

No.	α (1950.0)	δ	m_r	
			m_1	m_2
201	17ʰ37ᵐ13	−24°06′.44	15.9	16.5
202	37.13	−24 06.58	16.3	>17
203	37.24	−24 00.18	16.3	16.7
204	37.36	−24 15.70	15.8	16.4
205	37.40	−23 43.65	16.3	16.9
206	37.54	−23 49.86	16.1	16.7
207	37.62	−23 50.74	15.9	16.6
208	37.65	−23 31.78	15.7	16.4
209	37.65	−24 11.19	13.9	14.5
210	37.66	−23 57.38	14.0	>17
211	17 37.73	−23 30.73	15.6	16.4
212	37.87	−23 28.18	15.8	16.4
213	37.95	−23 43.40	15.2	15.9
214	37.95	−24 07.76	16.3	>17
215	38.15	−23 45.04	15.1	15.7
216	38.17	−23 45.01	15.8	16.5
217	38.17	−23 54.74	15.5	16.4

3. Discussion

(1) Each star is considered 'variable' if it is detected with the blink microscope and if it shows a $\Delta m = m_2 - m_1 \geq 0.4$ mag. Many other stars in the field are suspected variables ($\Delta m < 0.4$ mag.).

(2) Variable stars No. 41, 157 and 164 (Figure 3) seem to belong to the cluster. The study of their periods will give us a better idea as to their membership.

(3) Several of these 217 (102+115) variable stars found near NGC 6401 are probably longperiod or irregular variables.

The latest photographic observations dating only from June 1972, the determination of light curves, the calculation of periods, and the discussion of results, will form the subject of another publication.

4. Conclusion

These results show how profitable is the search for variable stars in the field of NGC 6401. But for the detection of variable stars within the cluster itself, it is absolutely necessary to obtain a great many photographic observations with an image tube at the Cassegrain focus of a reflector situated in the Southern Hemisphere.

On the other hand, it would be particularly interesting to proceed to a *UBV* photometric study of the cluster in order to study the $V-(B-V)$ colour-magnitude diagram.

We hope to accomplish this at the European Southern Observatory in April–May 1973.

References

Alter, G., Ruprecht, J. and Vanýsek, J.: 1970, *Catalogue of Star Clusters and Associations*, Akademiai Kiado, Budapest.
Arp, H. C.: 1965, *Stars and Stellar Systems* **5**, 401.
Herschel, W.: 1786, *Phil. Trans. Roy. Soc. London* **A76**, 457.
Hogg, H. B. S.: 1963, *Publ. David Dunlap Obs.*, No. 12.
Kron, G. E. and Mayall, N. U.: 1960, *Astron. J.* **65**, 581.
Kukarkin, B. V.: 1971, Private communication.
Terzan, A. and Rutily, B.: 1972. *Astron. Astrophys.* **16**, 408.

DISCUSSION

Demarque: What kind of image tube did you use?
Terzan: It will be described in the paper by Terzan, Rutily, and Ounnas on NGC 4590.

VARIABLE STARS IN NGC 4590 *

A. TERZAN and B. RUTILY

Lyons Observatory, European Southern Observatory, France

and

CH. OUNNAS

Nice Observatory, France

Abstract. The globular cluster NGC 4590 has been observed at Haute Provence Observatory (photographic photometry, 80-cm reflector, $F/6$) and at the European Southern Observatory (image-tube, 152-cm reflector, $F/15$).
- A sequence of red magnitudes, m_r ($\lambda_{\text{eff}} \cong 6400$ Å) has been established around the cluster. It comprises 28 stars with magnitudes between 11.66 and 16.50.
- Five new variable stars were discovered in a one square degree field, centered on the cluster.
- Four other variable stars (No. 39, 40, 41, and 42) were discovered in the cluster.

1. Introduction

According to Hogg (1972), 38 variable stars have been discovered in NGC 4590 ($\alpha = 12^h 36^m 8$; $\delta = -26°28'$, 1950) (Shapley, 1919; Hogg, 1939; Rosino and Pietra, 1954; Hogg, 1955; van Agt and Oosterhoff, 1957).

All these variables are RR Lyrae stars (Hogg, 1972), except for FI Hya, which is a field star of type M ($\max_{pg} = 10.2$, $\min_{pg} = 17.4$; $P = 324.1$ d., Sp = M4e).

In 1968 we began photographic observations of NGC 4590 to search for new variable stars.

2. Observations and Measures

I. HAUTE PROVENCE OBSERVATORY OBSERVATIONS

The first photographic observations were begun in 1968 (A.T.) and these were continued in 1970 (A.T., Ch.O.) and 1972 (A.T., B.R.) at the newtonian focus of the 80-cm reflector ($F/6$) in Haute Provence Observatory. These observations were made in the red wavelength range using Kodak 103 aE Plate + RG 1 (2 mm) filter, $\lambda_{\text{eff}} \cong 6400$Å.

First, we established a red photographic standard sequence in the field of the cluster, and then we searched for new variable stars in the cluster as well as in the surrounding field.

(a) *Red Sequence*

We have established a new standard red sequence by photometric transfer from SA 61 to the cluster field.

In SA 61, we have used the red magnitudes measured by Yoss (1955). Yoss's red wavelength range is centered on 6200 Å, which is very similar to ours (6400 Å).

* Observations made in part at Haute Provence Observatory (C.N.R.S.).

The transfer from one field to the other was made by using 7 independent photometric pairs. We have adopted the process described in a preceding study by Terzan (1965). The new standard m_r sequence in the field of NGC 4590 contains 28 stars in

TABLE I

Sequence of m_r magnitudes near the cluster NGC 4590

No.	m_r	No.	m_r	No.	m_r	No.	m_r
1	11.66:	8	14.00	15	15.04	22	15.94
2	12.36	9	14.02	16	15.16	23	15.99
3	13.10	10	14.24	17	15.24	24	16.14
4	13.29	11	14.38	18	15.38	25	16.27
5	13.51	12	14.56	19	15.51	26	16.36
6	13.70	13	14.66	20	15.66	27	16.39
7	13.85	14	14.86	21	15.80	28	16.50:

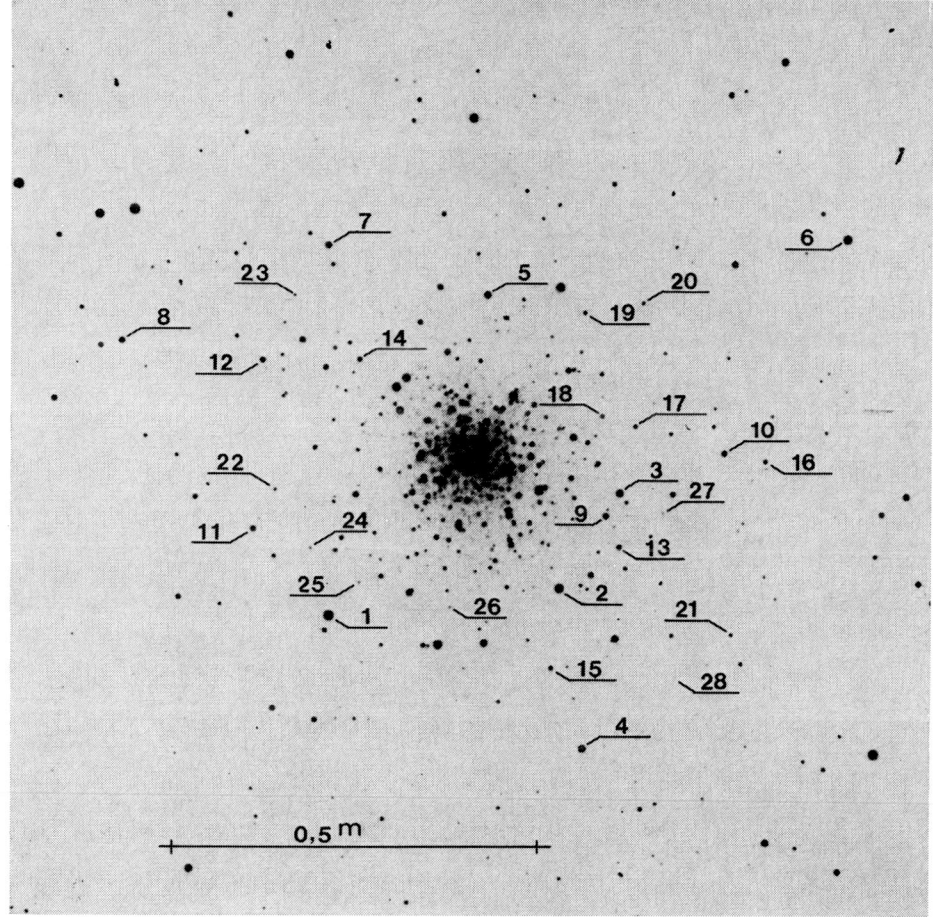

Fig. 1. Identification chart of new standard m_r sequence in the field of NGC 4590.

TABLE II

Positions and magnitudes of five new variables

No.	α	δ	m_r	
			m_1	m_2
1	$12^h36^m01^s$	$-26°52'11''$	14.8	15.5
2	12 36 19	-26 43 53	15.0	15.4
3	12 36 23	-26 10 40	14.4	14.9
4	12 38 32	-26 22 51	14.6	15.3
5	12 38 36	-26 04 10	15.0	15.7

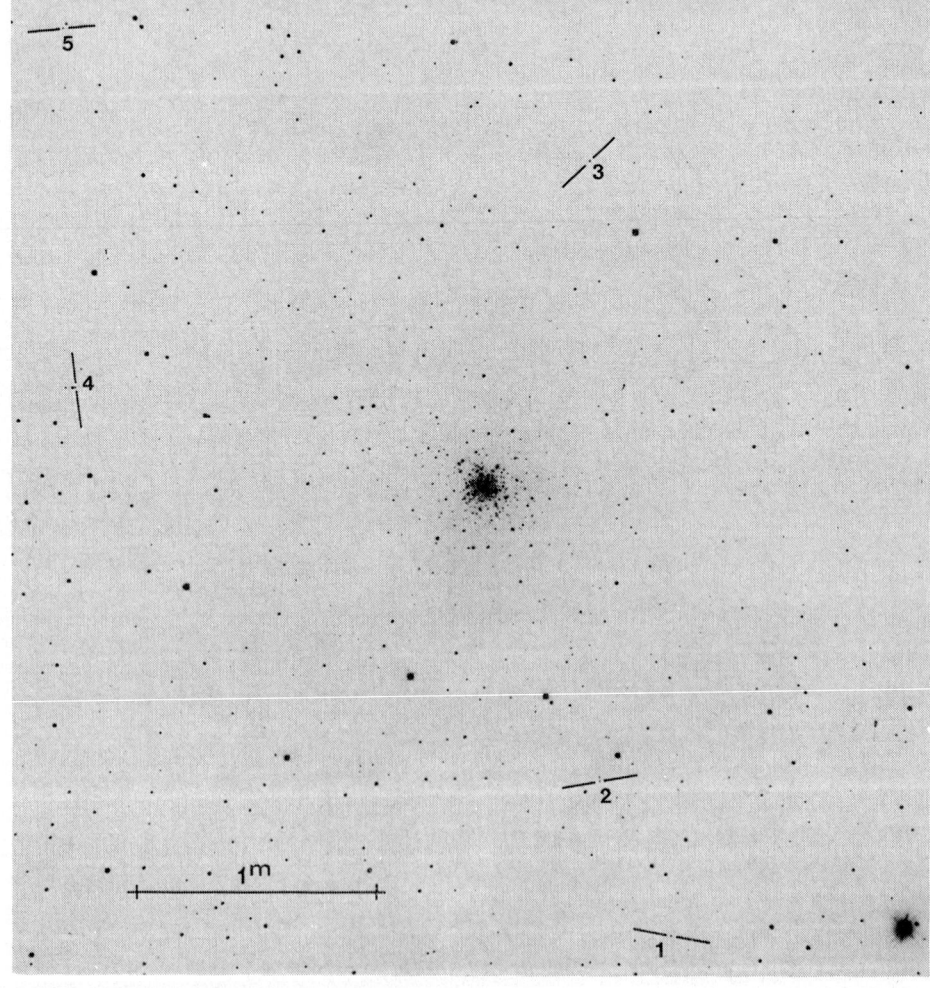

Fig. 2. Identification chart of 5 new variable stars detected near the globular cluster NGC 4590.

the magnitude range 11.66–16.50. The accuracy of our measurements is about ±0.04 mag. (Table I, Figure 1).

(b) *Search for new Variable Stars*

The search for new variable stars has been made with the blink microscope of the Lyons Observatory. We have not been able to study the central part of the cluster because the resolution on the plates is not sufficient. On the other hand, we have detected 5 new variable stars in a one square degree field centered on the cluster (Figure 2).

We have determined the magnitudes of these five variable stars from 25 plates. Table II contains the equatorial coordinates and the extreme magnitudes m_{r1} and m_{r2} of these stars.

We intend to carry out further studies of these stars to obtain light curves, periods, amplitudes and types of variability.

II. European Southern Observatory Observations

In 1972, Terzan made new observations of the cluster with an image tube, attached at the Cassegrain focus of the ESO 152-cm reflector ($F/15$). The image tube characteristics are as follows: ITT type, $F - 4708$, single stage, extended red, S-20. The field observed is circular with a diameter of 6.2 (40 mm).

We obtained two series of plates in the wavelength ranges B and IR:

B = image tube + BG 12 (2 mm) filter, $\lambda_{\text{eff}} \cong 4200$ Å.
IR = image tube + RG 695 (2 mm) filter, $\lambda_{\text{eff}} \cong 7600$ Å.

The observations were made in 1972 (April 14, 15, 17, 18 and 19).
The exposures were less than 20 min in B and 10 min in IR.

With these plates, we have been able to detect 4 more variable stars that are cluster members.

These stars have been numbered 39, 40, 41 and 42 (Figure 3) to add to the list of previously discovered variable stars in this cluster. In Table III, we give the rectangular coordinates (in seconds of arc) calculated by a method used by Hogg in the *Catalogue of Variable Stars in Globular Clusters*.

A preliminary study suggests that these stars are short period ($P < 1$d) and small amplitude (< 1 mag.) variable stars.

We suspect five other possible variables in the cluster, and new observations will enable us to check this.

3. Discussion

These preliminary results display the efficiency of our image tube in the search for variable stars in globular clusters.

Although the plates do not have good photometric quality, they permit us to determine light curves using comparative photometry between the variable star and and two other stars of the cluster.

These two comparison stars must be chosen near the variable star and must have a magnitude difference equal to or greater than the amplitude of the variable star.

Moreover, the mean exposure time of each plate is only about 20 min in B and 10 min in IR. Therefore we can get a large number of plates on each night and detect very short period variable stars.

TABLE III

Rectangular coordinates of four new variables

No.	x''	y''
39	-50	-8
40	-1	-52
41	$+4$	$+80$
42	-3	$+37$

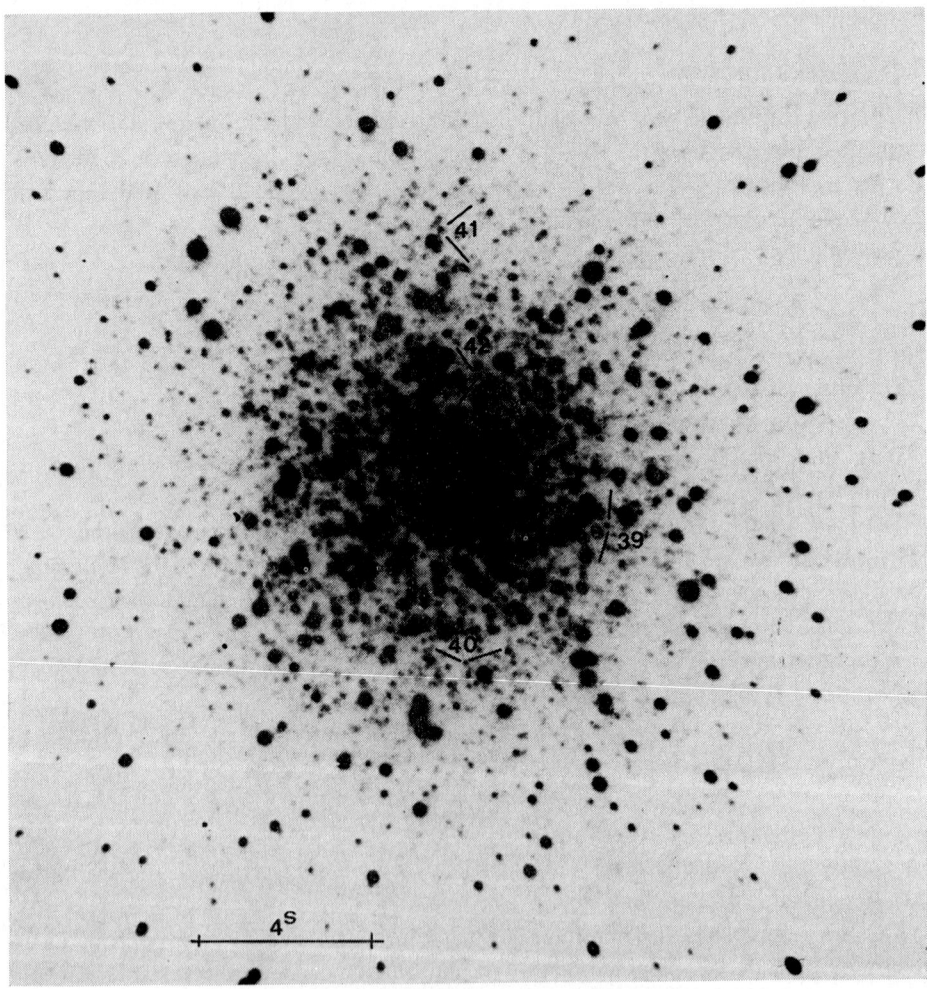

Fig. 3. Identification chart of 4 more variable stars detected in NGC 4590.

Further observations will permit us to continue the search for and the study of new variable stars in NGC 4590, as well as variables in other globular clusters situated or seen projected near the galactic center.

Acknowledgements

A. Terzan thanks the scientific Council of ESO and Prof. B. E. Westerlund, Director of ESO in Chili, for having given him a detached service (April 1972), and for having put an image tube at his disposal.

References

Agt, S. L. Th. J. van and Oosterhoff, P. Th.: 1959, *Ann. Sterrew. Leiden*, **21**, 253.
Hogg, H. S.: 1972.: *Publ. David Dunlap Obs.* **3**, No. 6
Rosino, L. and Pietra, S.: 1954, *Publ. Oss. astr. Univ. Bologna*, **6**, No. 5.
Sawyer, H. B.: 1939, *Publ. David Dunlap Obs.* **1**, No. 4
Sawyer, H. B.: 1955, *Publ. David Dunlap Obs.* **2**, No. 2
Shapley, H.: 1919, *Publ. Astron. Soc. Pacific* **31**, 266.
Terzan, A.: 1965, *Ann. Astrophys.* **28**, 935.
Yoss, K. M.: 1955, *Astron. J.* **60**, 338.

DISCUSSION

Cox: Have you found only nine variables so far?
Terzan: Five other new 'suspected' variable stars have also been detected in the cluster.

TWO-COLOUR PHOTOMETRY OF RR LYRAE VARIABLES IN NGC 6981

R. J. DICKENS and ROSALIND FLINN

Royal Greenwich Observatory, England

Abstract. NGC 6981 = M72 is a globular cluster with a late-type integrated spectrum (G3) and an ultraviolet excess of $\delta(U-B) = 0.15 \pm 0.05$, which places the cluster in the class of intermediate-to-high metal abundance clusters such as M3. The colour-magnitude diagram shows a horizontal branch populated on both sides of the variable star region. Previous work on the periods of the variables has indicated that the cluster is of Oosterhoff type I, with $\overline{P_{ab}} \approx 0.55$ day. Some results from B and V photometry of 21 RR Lyrae variables are described. Original measurements of about 50 candidate variables were made, including some possible new variables found by blinking plate pairs, but crowding and background photometric problems reduced the number of variables for which reasonable quality data could be obtained with the available 100-in plate material.

A new red variable near the centre of the cluster was discovered in the course of the work.

There appear to be relatively few c-type variables in the cluster, in spite of some uncertainties regarding undiscovered small-amplitude variables near the cluster centre. The sample discussed contains only one c type. The light and colour variations are derived and correlations between the light-curve parameters are discussed.

In particular the period-amplitude relation is similar to those of the clusters M3 and NGC 6171, both relatively metal-rich and also of Oosterhoff type I. The colour-magnitude diagram shows that there is no overlap in colour between ab and c-type variables, or between variable and non-variable stars. The period-colour diagram is used to derive physical parameters for the variables, using both linear and non-linear pulsation theory. The results essentially confirm earlier work on variables in other globular clusters in yielding a mean mass $\sim 0.5\ M_\odot$ or less and a helium abundance $\sim 30\%$, but the considerable uncertainties in these results, following work by van Albada and Baker (1971) are emphasized. A mean mass-to-light ratio of $\log\{(\mathfrak{M}/\mathfrak{M}_\odot)/(L/L_\odot)\} = -1.93$ is derived for the four clusters M3, ω Cen, NGC 6171 and NGC 6981 from their period-colour relations.

Details of this work are given in Dickens and Flinn (1972).

References

Albada, T. S. van and Baker, N. H.: 1971, *Astrophys. J.* **169**, 311.
Dickens, R. J. and Flinn, R.: 1972, *Monthly Notices Roy. Astron. Soc.* **158**, 99.

DISCUSSION

Baker: In clusters like this one, in which the number ratio of type-*ab* stars to type-*c* stars is very large, one would like to know more about possible selection effects which may create a bias against the discovery of type-*c* stars. An even better example is M3, which has a very small relative number of type-*c* stars, even for an Oosterhoff group I cluster. The number of stars per period interval for the type-*c* stars is much smaller than that for type-*ab* stars. This is odd, since the horizontal branch of M3 is well populated on both sides outside the variable strip. If there really is a gap on the HB in the region of the type-*c* variables, it might have very interesting implications for the evolution theory of HB stars.

Dickens: To be sure about the completeness, one should make photometric measurements of all stars in a given region of a cluster (or at least those in the relevant magnitude intervals). In most, if not in all clusters, this has not been done, so some incompleteness, at least among the smaller amplitude *c*-type variables, could well be present.

INTERMEDIATE BAND PHOTOMETRY OF RR LYRAE VARIABLES IN ω CEN AND 47 TUC

D. H. P. JONES

Mount Stromlo and Siding Spring Observatories, Research School of Physical Sciences, Australian National University, Australia

I have recently prepared for publication a study of 100 field RR Lyraes which were measured on the u, k, b, y, β system described by Jones (1971). It is shown that the following indices

$$\beta_1 = \beta - 0.125(k-b) + 0.103(b-y)$$
$$(u-b)_3 = (u-b) - 0.93(b-y) - 0.62(k-b) - 1.95(\beta - 2.75)$$
$$(k-b)_2 = (k-b) - 0.826(b-y) + 0.25(\beta - 2.75)$$

correlate strongly with θ_{eff}, $\log g$ and [Fe/H] respectively. Figures in that paper show the tightness of the correlations.

Six RR_{ab} variables were observed in ω Cen: Nos. 7, 33, 46, 74, 79 and 125 (twice) on the numbering system of Martin (1938). A 12 arcsecond aperture was always used. The seven observations were treated together as observations of one variable with the following results:

θ_{eff} $= 0.78 \pm 0.02$
$\log g$ $= 2.65 \pm 0.33$
[Fe/H] $= -1.45 \pm 0.34$
$\log(\mathfrak{M}/L) = -1.97 \pm 0.33$ in solar units.

Sargent (1965) analysed Fehrenbach's F giant in ω Cen and found [Fe/H] $= -1.59 \pm \pm 0.19$ when abundance anomalies among the different elements are ignored. The metal abundance is also in good accord with Kinman's (1959) classification of the red giants as type B (intermediate line weakening). The mass-to-luminosity ratio is in good accord with that found by Newell *et al.* (1969) for the horizontal branch stars. The luminosities of these variables may be estimated in two ways. They may be identified with the field halo RR Lyraes discussed by Clube and Jones (1971), who found $\langle M_v \rangle = +0.9$. A suggested downward revision to 1.3 has been severely criticized by Aslan (1971) and so is not considered here. With the above $\mathfrak{M}/L$ ratio, $M_v = +0.9$ corresponds to $\log \mathfrak{M} = -0.49$. Alternatively Christy's (1966) relation between luminosity and transition period from fundamental to first overtone indicates $\langle M_v \rangle = +0.6$; correspondingly $\log \mathfrak{M} = -0.31$.

From the $(b-y)$, β, plot a colour excess $E(b-y) = 0.09 \pm 0.04$ was derived. The blue horizontal branch stars are expected to respect the $k-b$, $b-y$ relation of the Population I stars, for at their temperature the K line vanishes whatever the abundance. Observations of two blue horizontal branch stars in ω Cen require a shift of $E(b-y)$

= 0.10 ± 0.02 to bring them on to the Population I relation. This reddening corresponds to $E(B-V) = 0.14$. If $\langle M_v \rangle = 0.8$ is accepted for the RR Lyraes then the true distance modulus of ω Cen is 13.4.

In the field of 47 Tuc there are only three RR Lyraes (Table I), two of which have been studied by Feast *et al.* (1960). Figure 1 gives the light curve of HV 810 which lies closest to the cluster. A 12 arcsecond aperture was always used, and because of the

TABLE I

RR Lyraes in the field of 47 Tuc

Variable	Bailey type	P (days)	Distance from 47 Tuc	$\langle y \rangle$	[Fe/H]
HV 809 (UX Tuc)	a	0.509	55′.2	13.97	−1.3 ± 0.5
HV 810	a	0.735	2.2	13.43	0 ± 0.4
HV 814	c	0.371	21.6	13.90	−1.2 ± 0.3

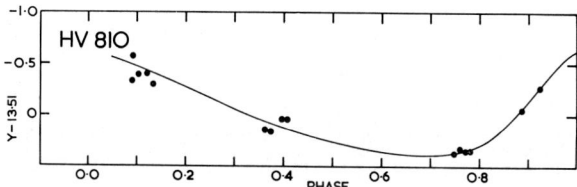

Fig. 1. Light curve of HV 810 as a function of phase $\phi = 1.36110$ (Helio JD-2440444.272).

rarity of good seeing at this zenith distance, the light curve required four observing seasons to complete. The Radcliffe period required a slight revision of approximately one cycle a year so that $P^{-1} = 1.36110$ for Figure 1. In Figure 2 the three variables are plotted on the colour magnitude array of Eggen (1972). HV 810 lies about 0.7 mag. above the horizontal branch which raises the possibility that it is affected by cluster background or an unseen companion. However this variable has an amplitude Δy of one magnitude which places it on the upper envelope of the period amplitude plot for field RR Lyraes. Correction for a companion would place it beyond the other variables. It is widely accepted that the period amplitude plot is an indicator of metal abundance with the highest amplitude variables being the most metal weak when the period is around 0.7 day. The solar metal abundance of HV 810, found from narrow band photometry is an exception to this rule. If HV 810 were given a low amplitude more accordant with its metal abundance then it would be necessary to postulate an unseen star at the sky position used: consequently the difference in brightness between the variables and the horizontal branch would be even greater.

While the other two variables have magnitudes closer to the horizontal branch, their metal abundances are substantially lower. HV 810 has an exceptionally high metal abundance for an RR Lyrae variable, and in view of the fact that it lies so close to one of the metal richest globular clusters known, it seems safe to accept it as

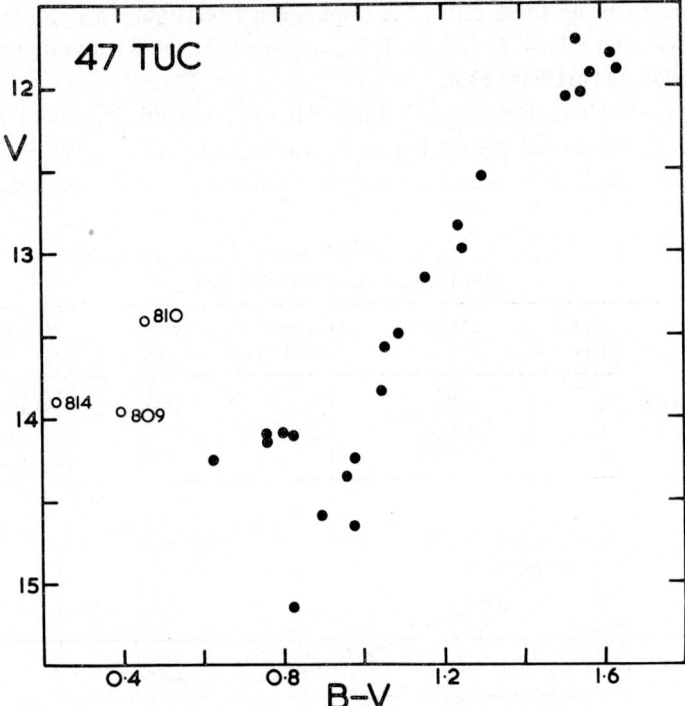

Fig. 2. Colour magnitude array of 47 Tuc. Filled circles are constant stars measured by Eggen. Open circles are the variables discussed in this paper identified by their HV numbers.

a member. The narrow band photometry gives for this star:

$$\theta_{\text{eff}} = 0.78 \pm 0.01$$
$$\log g = 2.71 \pm 0.23$$
$$\log(\mathfrak{M}/L) = -1.92 \pm 0.23$$

This star seems so exceptional that at present there is no meaningful way to estimate the luminosity. The other two variables are too weak in metals and too far from the cluster to be members.

References

Aslan, Z.: 1971, *Observatory* **91**, 14.
Christy, R. F.: 1966, *Astrophys. J.* **144**, 108.
Clube, S. V. M. and Jones, D. H. P.: 1971, *Monthly Notices Roy. Astron. Soc.* **151**, 231.
Eggen, O. J.: 1972, *Astrophys. J.* **172**, 639.
Feast, M. W., Thackeray, A. D., and Wesselink, A. J.: 1960, *Monthly Notices Roy. Astron. Soc.* **120**, 65.
Jones, D. H. P.: 1971, *Monthly Notices Roy. Astron. Soc.* **154**, 79.
Kinman, T. D.: 1959, *Monthly Notices Roy. Astron. Soc.* **119**, 538.
Martin, W. Ch.: 1938, *Ann. Sterrew. Leiden* **17**, (2).
Newell, E. B., Rodgers, A. W., and Searle, L.: 1969, *Astrophys. J.* **158**, 699.
Sargent, W. L. W.: 1965, *Observatory* **85**, 116.

DISCUSSION

Feast: Could you remind us of the proper motion position for the RR Lyraes in the region of 47 Tuc.

Jones: I believe Murray's observations indicate that HV 814 is a non-member. HV 810 had an unreliable proper motion because of crowding.

Demarque: What is your error estimate on the mass to light ratio?

Jones: A factor of two.

Schwarzschild: (a) May I ask at what phase of the light curve the observed colors are taken? (b) Is the calculated $\log g$ the same then as the 'static' $\log g$?

Jones: (a) Where $y = \langle y \rangle$ on the falling branch; (b) $\log g$ is the apparent gravity.

van den Bergh: What would main sequence fitting give for the absolute magnitude of HV 810?

Jones: Recent work by other people at Mt. Stromlo does not confirm the colours derived by Tifft from photographic transfers. Until this work is complete the question remains open.

UBV MAGNITUDES AND COLOURS OF 62 RR LYRAE STARS IN ω CENTAURI (NGC 5139)

E. H. GEYER

Astron. Institut der Universität Bonn, Observatorium Hoher List, Bonn, Germany

1. Introduction

It has been shown by the author (Geyer, 1967), and independently by Dickens and Woolley (1967), that there occur horizontal branch stars well within the RR Lyrae variable gap of the $c-m$-diagram of ω Centauri. Variability of these objects is not known, and presumably is unlikely, because the cluster is so well studied for variable stars. For the observationally best studied globular clusters M3, M13, M15 and M92 (Sandage, 1969), this phenomenon is *not* found. The author's original published $c-m$-diagram of NGC 5139 showed about 10 stars with $(B-V) \geq 0.2$ and V-magnitudes comparable with those of RR Lyrae stars of the cluster. Unpublished photoelectric measurements of some of these stars, carried out in 1968 with the 40-in photometric reflector of ESO by the author, support this finding. Also the study of blue horizontal branch stars by Newell, Rodgers and Searle (1969) of the Herstmonceux catalog of ω Centauri (Woolley *et al.*, 1966) confirms these results, the importance of which for the understanding of the horizontal branch stars and RR Lyrae variables is obvious.

2. Observations

UBV observations of larger samples of RR Lyrae stars in globular clusters have been published by Sandage (1959) for M3 and by Dickens (1970) for NGC 6171. Both clusters belong to Oosterhoff's 'short period group', whereas ω Cen is typical of the 'long period group'. As far as the morphology of the $c-m$-diagrams of globular clusters is concerned, ω Cen is between the M15/M92- and M10/M13-groups. For NGC 5139 intensity mean B, V values and colours based on photographic photometry for a larger number of RR Lyrae variables have been given independently by Dickens and Saunders (1965), and Geyer and Szeidl (1965, 1970). The latter observations are based on 38 plates in each colour obtained with the ADH-Baker-Schmidt camera of the Boyden Observatory, and on photoelectric standard stars which were also used for the *UBV*-photometry of the cluster (Geyer, 1967). These standards were also rechecked in 1968 at the European Southern Observatory. Several observers (see Newell *et al.*, 1969) have indicated that there exist systematic errors in the Herstmonceux photometry of NGC 5139, and these may also be responsible for the differences between our variable star photometry and that of Dickens and Saunders. On these plates 18 additional previously neglected RR Lyrae stars have been measured, especially V 68 which is an RR_c star with the extremely long period of $0^d.534$. These 62 RR Lyrae variables have now also been observed on 27 U-plates obtained with the ADH-camera in 1962–63 and 1968. The estimated internal errors of the intensity mean magnitudes $V_d{}^*$ are $\pm 0\overset{m}{.}02$; for $(B-V)_d{}^*$ $\pm 0\overset{m}{.}03$; and $\pm 0\overset{m}{.}04$ for $(U-B)_d{}^*$.

3. Discussion

We now discuss the observations of the ω Cen variables in comparison with those of M3, and concentrate mainly on the $(U-B)_d$ colours. The $(B-V)_d$ colours have been discussed by Geyer and Szeidl (1965, 1970).

From the available data we derive for the ω Cen RR Lyrae variables the following mean values:

RR_{ab}-stars ($N=31$):

$$\bar{P}_{ab} = 0\overset{d}{.}684;\quad \bar{V}_d = 14\overset{m}{.}45,\ \overline{(B-V)}_d = 0\overset{m}{.}336,\ \overline{(U-B)}_d = 0\overset{m}{.}292.$$
$$\sigma = \pm 0\overset{d}{.}104;\ \sigma = \pm 0\overset{m}{.}15,\quad \sigma = \pm 0\overset{m}{.}053,\quad \sigma = \pm 0\overset{m}{.}081.$$

RR_c-stars ($N=29$):

$$\bar{P}_c = 0\overset{d}{.}379;\quad \bar{V}_d = 14\overset{m}{.}42,\ \overline{(B-V)}_d = 0\overset{m}{.}243,\ \overline{(U-B)}_d = 0\overset{m}{.}334.$$
$$\sigma = \pm 0\overset{d}{.}060;\ \sigma = \pm 0\overset{m}{.}13,\quad \sigma = \pm 0\overset{m}{.}052,\quad \sigma = \pm 0\overset{m}{.}068.$$

Two stars have been omitted because of uncertain photometry of the one (V 15) and the ultrashort period of the second (V 65), which is generally considered to be a foreground dwarf cepheid. These values are in good agreement with the photoelectric photometry of 10 RR Lyrae stars in ω Cen by Oosterhoff and Walraven (1966). The dispersions in magnitude and colour are due to photometric errors *and* intrinsic scatter.

A. THE $c-m$-DIAGRAMS

In the V_d, $(B-V)_d$ diagram the borders of the variable gap are practically identical on the red side for M3 and ω Cen, and only slightly shifted on the blue side for the ω Cen variables. Also both clusters have in common that the $(B-V)_d$ for RR_c stars is smaller than for the RR_{ab} variables, but only in ω Cen do both star types overlap in colour. Yet in the V_d, $(U-B)_d$ diagram the colour index is smaller for the RR_{ab} than for the RR_c stars. On the other hand the $(U-B)_d$ width of the variable gap is wider for NGC 5139 than for M3, and in addition its colour borders are shifted by up to $0\overset{m}{.}35$ to the red, as can be seen from Figure 1. The eclipsing variable V 78 (Geyer, 1971) is well situated within the RR Lyrae domain. Together with its photoelectrically derived colours this is an additional argument for its cluster-membership. V 92 is a $1\overset{d}{.}345$ Cepheid.

B. THE PERIOD-COLOUR RELATIONS

In the period- $(B-V)_d$ diagram of ω Cen the slope for the RR_{ab} stars is less steep than for the same variables in M3, whereas the slope for the RR_c stars is nearly

---

* V_d, $(B-V)_d$, $(U-B)_d$ are identical with $\langle V \rangle$, $\langle B \rangle - \langle V \rangle$, and $\langle U \rangle - \langle B \rangle$ respectively.

identical with that of M3. The previously suspected second RR_c branch parallel to the first one, but shifted to longer periods, has been confirmed by 7 additional RR_c stars, especially by V 68, which defines the upper end of this branch. This is of importance since it indicates that there is a mass range of the RR_c stars in ω Cen.

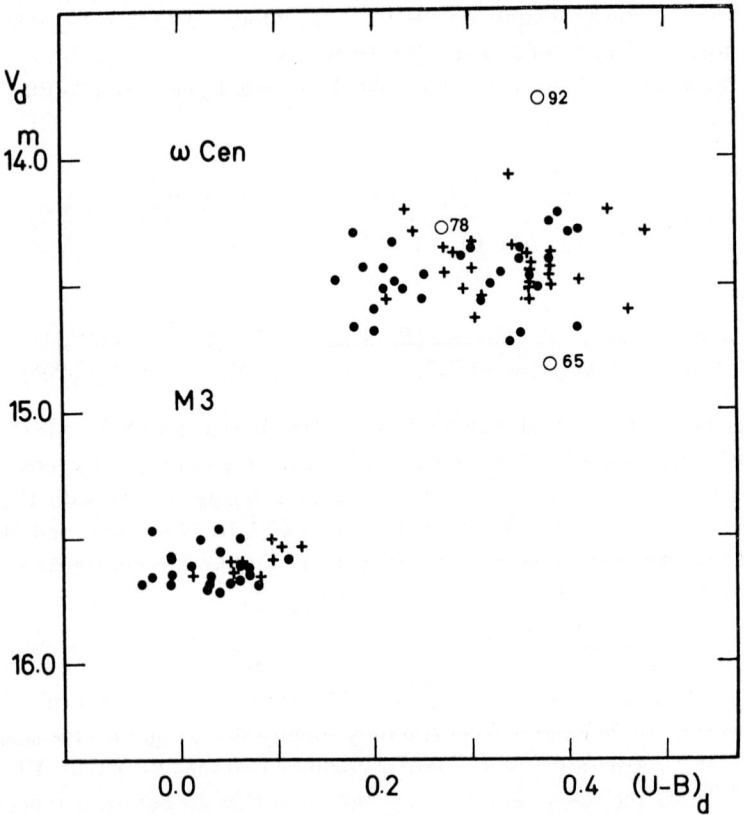

Fig. 1. The $(U-B)_d$-magnitude diagram for RR Lyrae variables of the globular clusters ω Cen and M3. Filled circles are RR_{ab} stars, crosses are RR_c stars. Star 65 is a $0\overset{d}{.}035$ dwarf cepheid; star 78 is an EA binary and star 92 a $1\overset{d}{.}34$ Population II-Cepheid.

As can be seen from the period- $\overline{(U-B)}_d$ plot (Figure 2) no obvious period-colour correlation exists.

C. THE PERIOD-LUMINOSITY RELATION

It has been shown previously by Dickens and Saunders (1965), and Geyer and Szeidl (1965, 1970) that there exist period-magnitude relations for the ω Cen RR Lyrae variables. The additional observed variables strengthen this result as can be seen

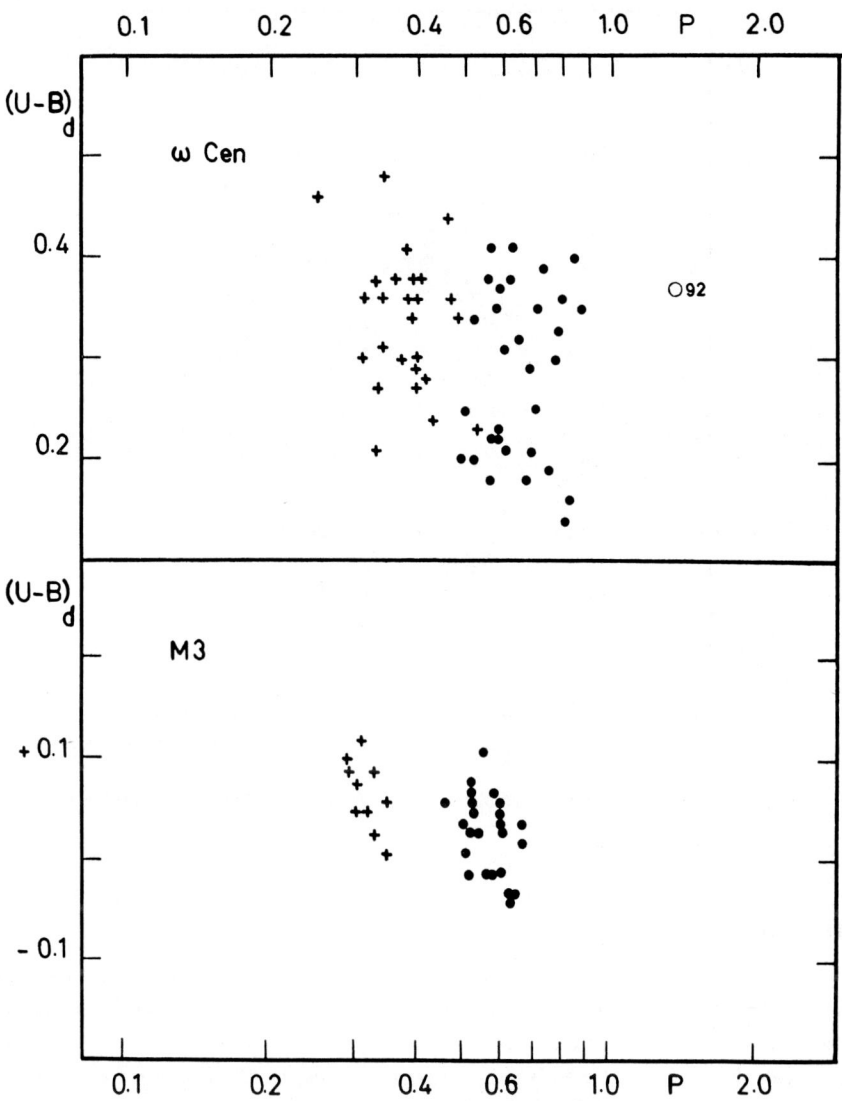

Fig. 2. The period-$(U-B)_d$ relations. Symbols as in Figure 1.

from the period-magnitude plot (Figure 3). The empirically derived relations are as follows:

ω Cen: RR$_{ab}$ stars:
$$V_d = -1.99 \cdot \log P_{ab} + 1.83 \cdot (B-V)_d + 13.45.$$

RR$_c$ stars:
$$V_d = -1.39 \cdot \log P_c - 0.13 \cdot (B-V)_d + 13.85.$$

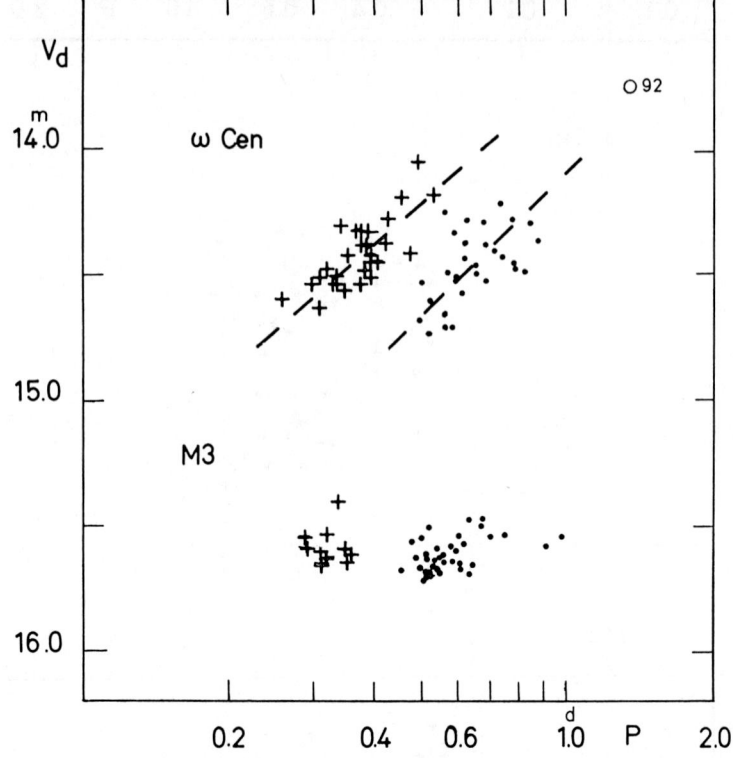

Fig. 3. The period-magnitude diagram for ω Cen and M3. Symbols as in Figure 1.

D. THE TWO-COLOUR DIAGRAM

In the two-colour plot (Figure 4) the M3-variables and horizontal branch stars show only a slight ultra-violet deficiency in comparison with the population main-sequence relation, and there is a quite smooth transition from non-variable to variable stars. This is also observed for the NGC 6171 stars (Dickens, 1970). The RR Lyrae variables of ω Cen behave quite differently, showing a *uv* deficiency amounting to $0^{m}\!.4$. Yet the majority of the blue horizontal branch stars which also overlap the variable gap in the $c-m$-diagram do *not* show such a large *uv*-excess. Only some of them coincide with the position of the RR Lyrae stars. They may therefore be candidates for variability. Actually one of them is the EA binary V 78, for which the author showed that the hotter, smaller and more luminous component of the system is intrinsically variable. It should be mentioned that a large *uv*-excess in some RR Lyrae field stars is reported by Sturch (1966) and recently by Stępień (1972).

We therefore come to the conclusion that the ω Cen RR Lyrae variables are in a quite different evolutionary stage, and show differences in their atmospheres compared to the adjacent horizontal branch stars.

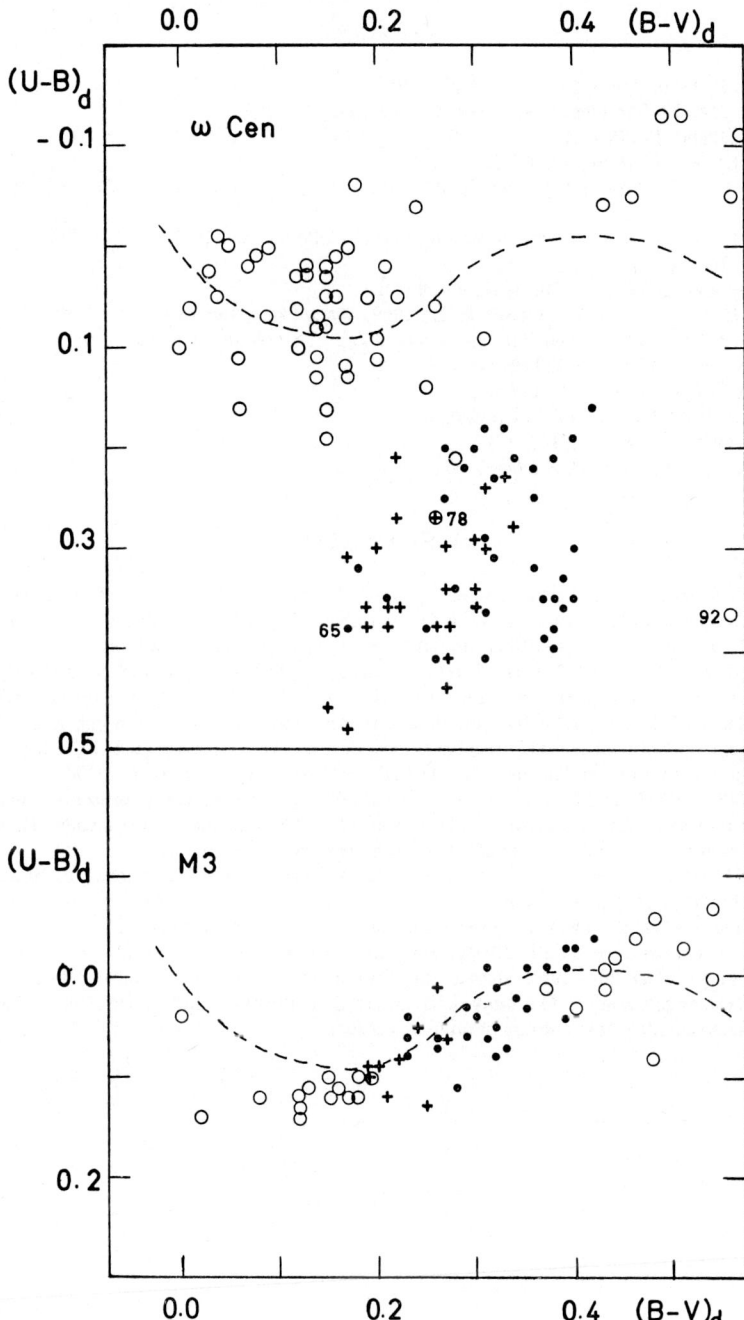

Fig. 4. The two-colour diagram for horizontal branch and RR Lyrae stars. Open circles are horizontal branch stars; other symbols are as in Figure 1.

References

Dickens, R. J.: 1970, *Astrophys. J. Suppl.* **22**, 249.
Dickens, R. J. and J. Saunders,: 1965, *Roy. Observ. Bull.*, No. 101.
Dickens, R. J. and Woolley, R. v. d. R.: 1967, *Roy. Observ. Bull.*, No. 128.
Geyer, E. H.: 1967, *Z. Astrophys.* **66**, 16.
Geyer, E. H.: 1971, Proceedings IAU Colloquium No. 15, *Veröff. Remeis-Sternw. Bamberg* **IX**, No. 100.
Geyer, E. H. and Szeidl, B.: 1965, Proceedings IAU Colloquium No. 3, *Kl. Veröff. Remeis-Sternw. Bamberg*, No. 40.
Geyer, E. H. and Szeidl, B.: 1970, *Astrophys.* **4**, 40.
Newell, E. B., Rodgers, A. W., and Searle, L.: 1969, *Astrophys. J.* **158**, 699.
Oosterhoff, P. Th. and Walraven, Th.: 1966, *Bull. Astron. Inst. Neth.* **18**, 387.
Sandage, A.: 1959, *Astrophys. J.* **129**, 596.
Sandage, A.: 1969, *Astrophys. J.* **157**, 515.
Stepień, K.: 1972, *Acta Astron.* (in press).
Sturch, C.: 1966, *Astrophys. J.* **143**, 774.
Woolley, R. v. d. R.: 1966, *Roy. Observ. Ann.*, No. 2.

DISCUSSION

Jones: Christy predicts a complete absence of variables between $3/4\ P_{tr}$ and P_{tr} ($P_{tr} = 0\overset{d}{.}565$ for ω Cen) but Dr Geyer's observations show a few variables in this range. Christy suggested that such stars might correspond to those of his models where both fundamental and first harmonic could be excited. Martin, in his Table II, lists 16 stars with irregular light curves in ω Cen, and 44 percent of them lie in Christy's gap. Of the remaining 109 RR Lyraes only 6 percent lie in the gap. While this is strong evidence of the simultaneous excitation of two modes, the boundary in period between types a and c is remarkably clear-cut. Of the a type, the shortest period with a regular light curve is No. 74 ($P = 0\overset{d}{.}503$) or with an irregular curve No. 112 ($P = 0\overset{d}{.}474$). The two longest period c type variables are No. 68 ($P = 0\overset{d}{.}534$) and No. 47 ($P = 0\overset{d}{.}485$). No. 68, of which special mention was made by Dr Geyer, has a proper motion indistinguishable from the other variables in the cluster. However it is 0.26 mag. brighter (m_{pg}) and may therefore not be a member.

Dickens: Mention was made in the paper of systematic error in the Herstmonceaux photometry of horizontal branch stars which was previously pointed out by the Stromlo workers. It should be noted that a comparison of the mean colours of variables in common between the work of Geyer and Szeidl and ours show a systematic difference in the run that Geyer and Szeidl's colours are in the mean 0.07 magn. *bluer* than ours, whereas those obtained at Stromlo for non-variables of a similar colour are somewhat *redder*. This remark is intended to illustrate and to emphasize the need for further work to decide whose colours (if any) are correct.

TWO-COLOR OBSERVATIONS OF RR LYRAE STARS IN M14

AMELIA WEHLAU and NICHOLAS POTTS
University of Western Ontario, London, Canada

Abstract. Two-color photographic photometry has been carried out on RR Lyrae variables in M14. The position of these variables in the color-magnitude diagram is shown, and it is noted that there is a lack of horizontal branch stars to the red of the variable star gap. Color-amplitude, period-amplitude and color-period relations for the RR Lyrae variables measured are shown.

M14 (NGC 6402, $\alpha = 17^h 35^m.0$, $\delta = -3°15'$, $l^{II} = 21°$, $b^{II} = +14°$) is a cluster of intermediate metal content, placed in class IV by Morgan (1959). It has a good number of variable stars, with 5 Population II Cepheids and more than 70 RR Lyrae variables, with periods determined for about 45 of the latter type (Sawyer Hogg and Wehlau, 1966, 1968). However it is not an easy cluster to study because it has a distance modulus of more than 15 magnitudes and in addition appears to have more than a magnitude absorption in the blue.

A series of 42 plates of this cluster was taken with the 48-in. telescope of the University of Western Ontario from 1970 to 1972 and, in addition, 11 plates taken by Dr S. Demers in 1967 with the Naval Observatory 61-in. at Flagstaff, and 23 plates taken in 1952 by Dr H. Arp with the 60 and 100-in. telescopes at Mt. Wilson, were measured. Altogether the plate material consists of 33 blue and 43 visual plates of the cluster.

Two separate photoelectric sequences obtained by Dr Demers and Dr R. Racine were combined and used as one sequence with limiting magnitudes of 18.8 in B and 17.8 in V There appears to be no systematic differences between the two sequences.

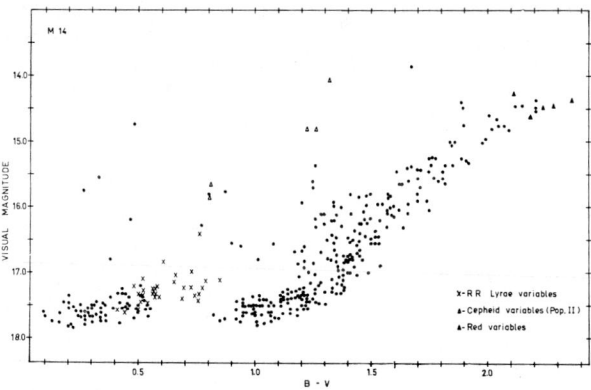

Fig. 1. Color-magnitude diagram for M14 showing the position of the variable stars. The nonvariable stars represent about four-tenths of the cluster area. All of the known Population II Cepheids and red variables are shown and about one-half of the known RR Lyrae variables. No stars below the limiting magnitudes of the sequence are shown.

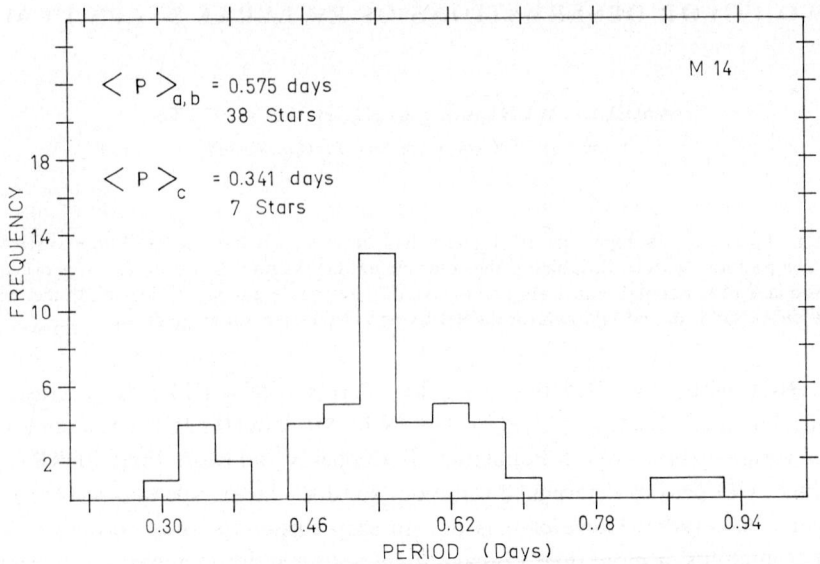

Fig. 2. Frequency of period histogram for RR Lyrae variables with known periods in M14.

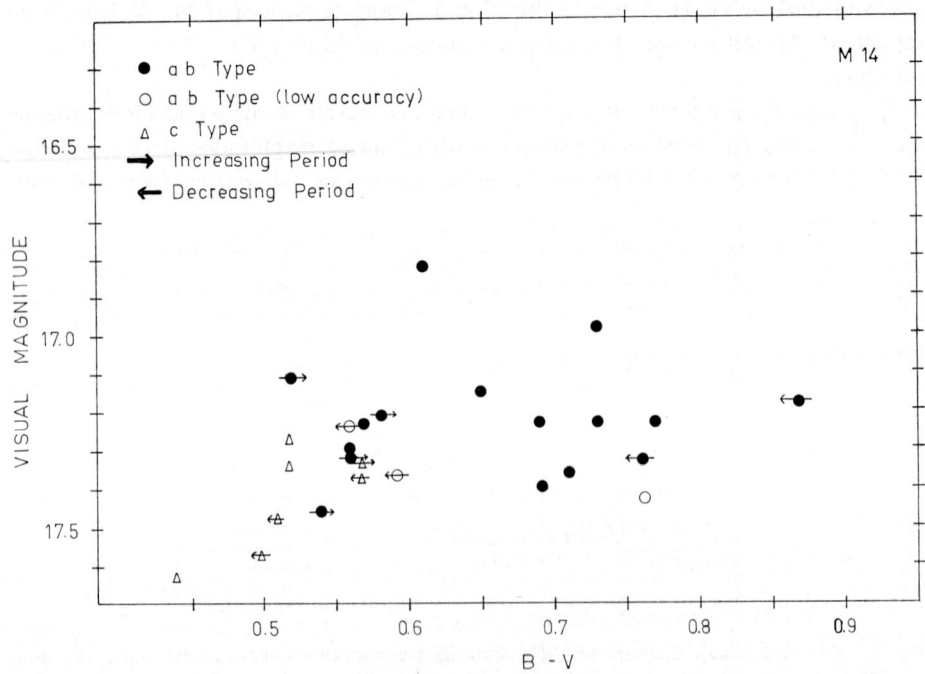

Fig. 3. The variable star region of the horizontal branch showing relation between color and period change.

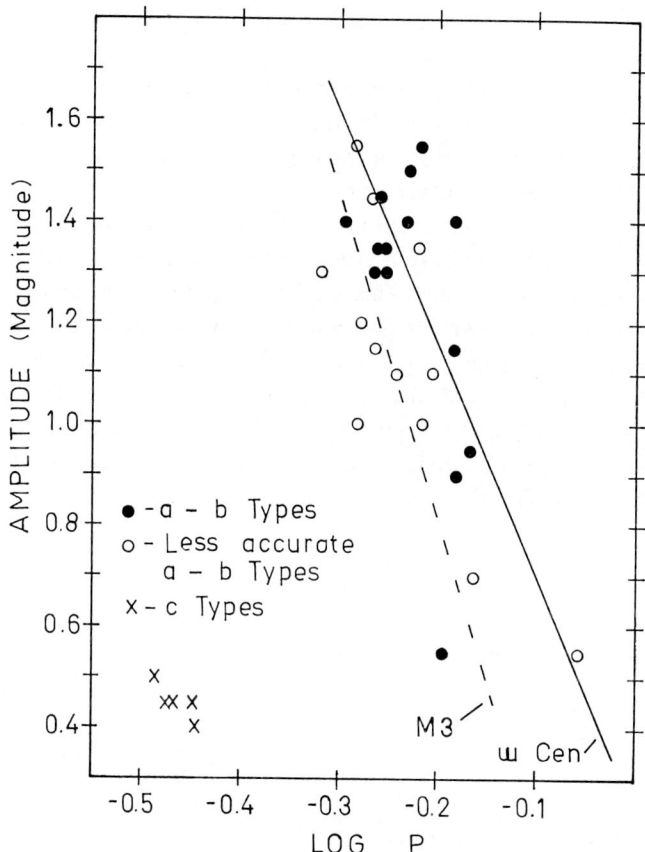

Fig. 4. Period-amplitude relation for RR Lyrae variables in M14. The positions of mean lines for variables in ω Cen and M3 are also shown.

Light curves were constructed in both colors for 25 *ab*-type stars. These were transformed into intensity curves and integrated in order to obtain mean colors. Periods are known for 7 *c*-type variables but we were unable to get satisfactory visual light curves for any of these. Being fainter visually than the *ab*-types they fall very close to the plate limit on the visual plates and in many cases there appeared to have been a period change between the sets of observations. However we did have enough observations in both colors to obtain somewhat less accurate values of $B-V$ for them as well as other *ab*-type variables.

Figure 1 shows a color-magnitude diagram for M14 recently determined by Christine Smith (1973), with the RR Lyrae variables included in the diagram. The non-variable stars represent somewhat less than half the globular cluster. Stars not known to be variable lying in the *ab*-type portion of the variable star gap were investigated and in almost all cases were found to be either a blend of more than one image or to be varying. In this way 11 new variables have been discovered. Such stars in the

c-type region have yet to be investigated. Because of their lower amplitudes their variability is more difficult to determine.

Because of this uncertainty as to which stars are variable the blue edge of the variable star gap is not accurately determined. There is an obvious lack of horizontal branch stars to the red of the gap. On the basis of metallicity and Oosterhoff group, such a lack would not have been expected.

A good value for the reddening was not available for this cluster, so we decided to use the method developed by Sturch (1966, 1967), which uses the colors of ab-type RR Lyrae stars during minimum. This method requires a knowledge of the ultraviolet excess which is unknown for this cluster. However there appears to be a relationship between the frequency distribution of RR Lyrae periods in a cluster and the ultraviolet excess. The frequency distribution for M14 is that of an Oosterhoff group I cluster and is shown in Figure 2. On the basis of this distribution a value of δ of 0.08 mag. was used to obtain a reddening of 0.35 mag. This is considerably lower than values obtained by others using the integrated color of the cluster.

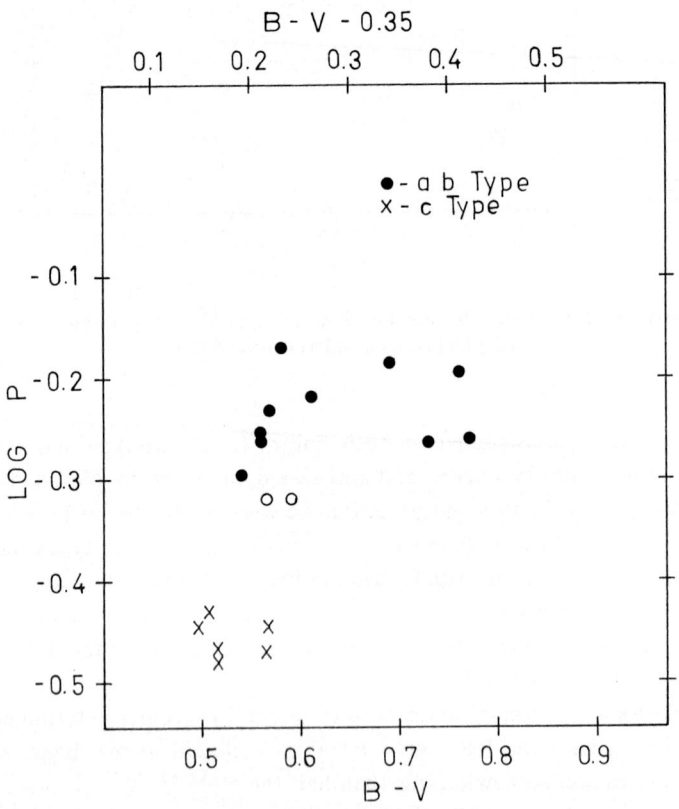

Fig. 5. Period-color relation for RR Lyrae variables with the most accurate colors in M14. The open circles are less accurate colors for variables 14 and 33 which have periods near 0.47, the transition period for this cluster.

Figure 3 shows the color-magnitude diagram for the RR Lyrae stars alone. The arrows show the sense of period change when such a change is observed. The observed changes appear to be sudden and random and not easily attributed to evolutionary changes.

A plot of period versus amplitude is shown in Figure 4. Mean lines for the variables in ω Cen (Dickens and Saunders, 1965) and M3 (Roberts and Sandage, 1955) are shown as well.

Figures 5 and 6 show the relationship between color and period and color and amplitude respectively. Only those *ab*-type variables which fall in completely uncrowded areas are plotted since colors obtained for the other variables appear to be reddened through proximity to red stars in the cluster.

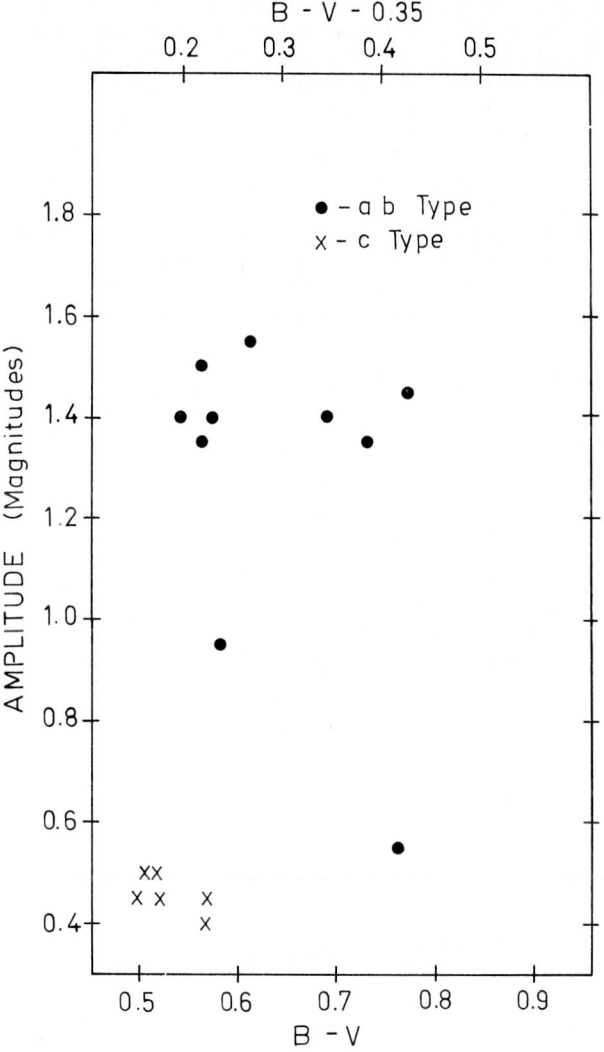

Fig. 6. Color-amplitude relation for RR Lyrae variables with the most accurate colors in M14.

References

Dickens, R. J. and Saunders, J.: 1965, *Roy. Obs. Bull.* No. 101.
Morgan, W. W.: 1959, *Astron. J.* **64**, 432.
Roberts, M. and Sandage, A.: 1955, *Astron. J.* **60**, 185.
Sawyer Hogg, H. and Wehlau, A.: 1966, *Publ. David Dunlap Obs.* **2**, No. 17.
Sawyer Hogg, H. and Wehlau, A.: 1968, *Publ. David Dunlap Obs.* **2**, No. 19.
Smith, C.: 1973, unpublished M.Sc. thesis, University of Western Ontario.
Sturch, C.: 1966, *Astrophys. J.* **143**, 774.
Sturch, C.: 1967, *Astrophys. J.* **148**, 477.

DISCUSSION

Cox: What fraction of the stars in the variable gap do you feel are non-variable?

Wehlau: The fraction where the *ab*-types are found is probably very low as the stars that were thought to be non-variable have been looked at and so far all have turned out to be new variables or blended with another star image. There are more such stars found in that part of the gap where the *c*-type variables appear. However they may also turn out to be variable as it is much harder to recognize the variability of these fainter stars with smaller amplitude of variation. We plan to make a greater effort to check the constancy of these stars.

Schwarzschild: (To Mrs Hogg and Mrs Wehlau). Do you have any comment concerning the earlier history of the search for variables in this cluster?

Hogg: I discovered 72 variables in this cluster from plates at the Dominion Astrophysical Observatory in the 1930's. Then in recent years Dr Amelia Wehlau has been collaborating with me on the the determination of periods. We have already published forty and have some others determined, but not published. Dr Wehlau discovered four more variables and one nova.

ON THE CHANGES OF PERIODS OF RR LYRAE VARIABLES IN THE GLOBULAR CLUSTERS M5, M53 AND NGC 5053

V. P. GORANSKIJ, B. V. KUKARKIN, and N. N. SAMUS'

Moscow University, Moscow, U.S.S.R.

The period of a variable star is very sensitive to extremely minor changes in luminosity, mass, radius, surface temperature, etc. of the star. With new theoretical calculations of the interior structure of stars in the late stages of evolution now available, the investigation of instabilities in the periods of the RR Lyrae variables in globular clusters becomes urgent.

Sufficient material for statistical investigation of the instability of periods of cluster variables has already been accumulated for several clusters. For example, the observations of the RR Lyrae variables in the globular clusters M3 and M5 embrace a time interval of 80 yr, in M53 – 50 yr, in NGC 5053 – 45 yr. Such time intervals afford the possibility of defining the periods of variables more accurately and obtaining information about their changes.

For the majority of variables in three globular clusters studied by us, long time intervals of relative quietness of the periods replaced by brief intervals of spontaneous changes are highly characteristic. Therefore during the intervals of quietness the residuals $O-C$ can be well approximated by linear elements, and the general $O-C$ diagram can be well approximated by discontinuous lines. Of the 51 variables studied in M5 (Kukarkin and Kukarkina, 1971) 85% show sudden changes of period. 93% of period-changing variables in M53 (of the 13 studied) and 75% of the period-changing variables in NGC 5053 (of the 9 studied) changed their periods spontaneously. In exceptional cases the residuals $O-C$ are subject to slight oscillations during the time intervals of quietness. In other cases the considerable dispersion of observations does not permit a conclusion about the character of period changes to be drawn. All the RR Lyrae variables with the Blazhko-effect show very large changes of periods.

Some interesting conceptions on the problem of period changes were published by Detre (1969).

Unfortunately, up to now the investigators of variable stars have supported the tradition of approximating the residuals $O-C$ by a quadratic form as follows:

$$O - C = T_0 + P \cdot E + k \cdot E^2$$

or by a trigonometric equation like

$$O - C = T_0 + P \cdot E + k_1 \cdot \sin[\theta \cdot (E - E_0)]$$

and to use the quantities k, k_1 and θ as measures of period instability. However, as long as 15 yr ago it was shown (Parenago, 1956) that these equations do not represent the observations if the investigator utilises an array of observations which is sufficiently prolonged. Modern theoretical calculations of stellar evolution predict smaller rates of secular period changes; the observed changes of periods must be a kind of noise.

Figure 1 shows several examples of unsuccessful interpretations of the $O-C$ curves by means of quadratic and trigonometric equations (Margoni, 1964, 1967; Wachmann, 1965) for 4 RR Lyrae variables in the globular cluster M53. One can see that usually the more complete array of observations contradicts such an interpretation.

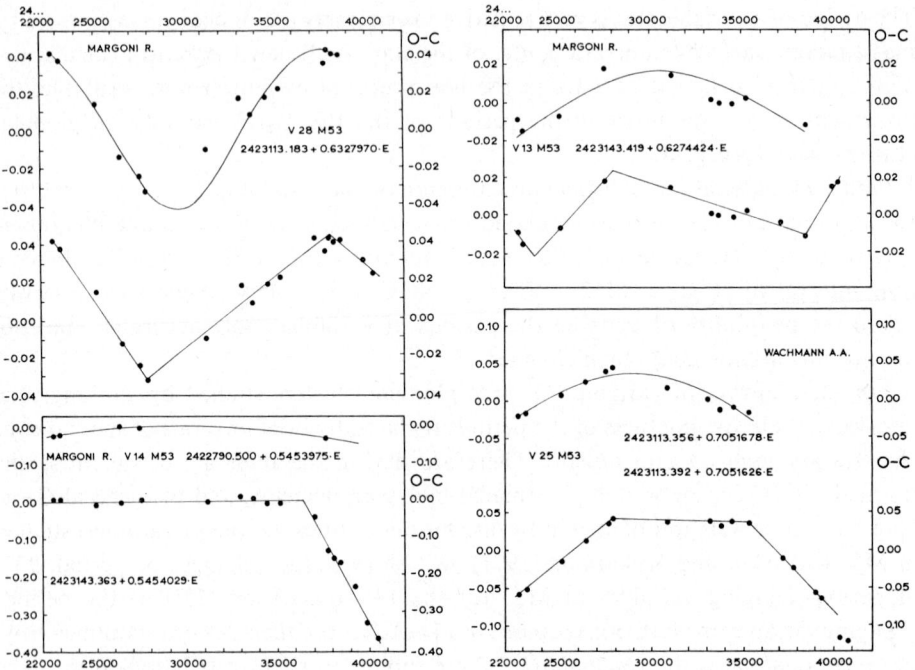

Fig. 1. Typical $O-C$ diagrams for RR Lyrae variables in globular clusters. Note that the discontinuous straight lines generally represent the observations better than the quadratic or sinusoidal curves.

We suppose that the best physical quantities describing the instability of the period of an RR Lyrae variable (and also other pulsating stars) and reflecting the nature of these period variations are the value of the period jump ΔP, the duration of the quiet stage of the period ΔT_1 before and ΔT_2 after the jump. It is very probable that the quantity $\log(\Delta T \cdot \Delta P/P)$ may be a measure of instability (Parenago, 1956). All these quantities should be used for statistical analysis.

References

Detre, L.: 1969, in *Colloquium on Variable Stars*, 4th, Budapest, 1968, *Non-Periodic Phenomena in Variable Stars*, L. Detre (ed.), Academic Press, Budapest, p. 3.
Kukarkin, B. V. and Kukarkina, N. P.: 1971, *Perem. Zvezdy*, Suppl. **1**, No. 1.

Margoni, R.: 1964, *Contr. Oss. Astrofis. Univ. Padova* No. 150.
Margoni, R.: 1967, *Contr. Oss. Astrofis. Univ. Padova* No. 198.
Parenago, P. P.: 1956, *Perem. Zvezdy* 11, 236.
Wachmann, A. A.: 1965, in *Astronomische Abhandlungen; Professor Dr Cuno Hoffmeister zum 70. Geburtstage gewidmet*, Barth, Leipzig, p. 121.

DISCUSSION

Schwarschild: I suspect that the question regarding the character of period changes – other than the probably very slow systematic evolutionary changes – is of basic importance to theoretical considerations. It would seem to me, for example, that it would make all the difference for theoreticians to know whether the random period changes are relatively abrupt, as suggested in the paper we just heard, or relatively smooth as implied by β computations.

Coutts: Some of the periodic period changes found by Margoni can be explained if the RR Lyraes are members of binary systems. The variation in radial velocity causes an apparent period change which can be detected if the period of the binary is at least 8 yr.

PERIOD AND PERIOD CHANGES OF
RR LYRAE VARIABLES IN M15

A. J. WESSELINK
Yale University, U.S.A.

Abstract. New observations of the variables in M15 have been made with the Yale 40-in. reflector for the study of periods and period changes.

Since the counting of cycles from the time of Bailey's observations to the present day cannot always be made unambiguously, there is a corresponding uncertainty in the nature and amount of the period changes. Good lightcurves are obtained even when the periods used for their reduction are incorrect because the number of cycles in a day, or in a month, or in a year, is wrong by one. When this happens, however, period changes are likely to be indicated, and these will be spurious.

Nevertheless, it can be shown that the periods of most variables have not been constant during the interval 1900–1970.

The situation could be improved if a relatively small number of plates were to be obtained each year for several decades, preferably with the same instrument. Individual accuracy could be improved by the use of the electronic camera.

The full contents of this paper will be published elsewhere.

DISCUSSION

Feast: What is the effect of variations in the light curve and shifting of time of maximum light?

Coutts: When computing phase-shift diagrams, often the observations from 2 or 3 yr are combined into one light curve. Thus, if the light curve varies from cycle to cycle, the light curve plotted with observations from more than one year will have scatter due to this. This causes the error in the determination of the points for the $O - C$ diagram.

Wesselink: I have never combined in a mean curve observations which I suspected of showing a change in the form of the light curve. So there was no problem in defining epochs of maximum light for $O - C$ diagrams for period change.

PERIOD CHANGES AS EVIDENCE
FOR EVOLUTION IN ω CEN

EMILIA PISANI BELSERENE

Dept. of Astronomy, Columbia University; Dept. of Physics and Astronomy,
Lehman College of the City University of New York, N.Y., U.S.A.

Much has been said at this colloquium to the effect that observed changes in the periods of variable stars are so fast or so abrupt or so quick to change sign that they cannot possibly be attributed to stellar evolution. If you are wondering how this report can dare to make a contradictory claim read the title carefully. It does not promise that there really is evidence for evolution in Omega Centauri. It only proposes to look at the period changes with evolution in mind. We may all decide that the evidence is strong or that it is weak or perhaps that there is no evidence at all.

The stars in this study are the 43 most regular and best observed of the RR_{ab} variables in the cluster (NGC 5139). As usual the test of the correctness or regularity of a period begins with the deviations, $O-C$, of observed times from predictions based on an assumed constant period. If the $O-C$ diagrams were reproduced here you would see that for some of the stars a parabola, i. e., a constant rate of change of period, is as good a representation of the observations as one could expect. You would also see several cases where the behavior is clearly too complex to be satisfied by so simple a curve as a parabola and others where a straightline, implying no change in period, is entirely satisfactory. But being able to get away with a constant period does not mean that the period has really been constant nor does the presence of clearly non-evolutionary behavior mean that the effect of evolution is necessarily absent. It is good to have an objectively assigned set of numbers among which to look for evidence of evolution, even though we recognize that the period may not have changed at all or, if it has, that a secular change is certainly not the whole story and perhaps is not even part of it. We can derive from the observations some sort of average, apparent rate of change. It is always possible to force a parabola through any sort of $O-C$ plot, after all, by the simple expedient of combining the epochs into three mean points.

This is what I have done and have found, as Martin (1938) already found in his extensive study of Omega Cen, that the rate of change of period – let us call it β as Martin did but decide to measure it in cycles per million years – is positive for some stars and negative for others but that positive values predominate. Before we look at the results in detail let us consider various processes which can affect the observed rate of change of period.

Including stellar evolution there are at least six reasons to expect observations at three epochs to show a deviation from a perfectly constant period. The one that comes to mind first is random observational error. The mean error of the magnitudes is typically about 0.05 mag. This leads to a mean error in the epochs of about 10 min, depending on the number of observations. The error in a mean period goes as the reciprocal of the time interval and the error in a value of β calculated from three epochs, i.e., two time intervals, goes as the inverse square, turning out to be of the order of one-tenth of a cycle per million years.

Next there may be some sort of cosmic mean error, caused by failure of the light curves to repeat exactly. They may actually be multiply periodic, as in the Blazhko effect, but they look irregular if the observations are at more or less random intervals. The effect on the light curves is mostly just scatter, very much like the effect of observational error. Stars with obviously irregular light curves have not been included in this study, but minor irregularities almost certainly occur and we can expect them to cause fluctuations of several minutes in the epochs and again errors of something like 0.1 cycle in β.

A third way to get a non-zero value of β is a different sort of random process. Is it possible that in the pulsation of a star the lengths of individual cycles deviate just slightly and randomly from a basic period (Sterne, 1934)? Fluctuations in period tend to accumulate, whereas fluctuations in epoch tend to cancel out. The effect on an average period goes as the inverse square root of the time interval instead of as the reciprocal. The effect on β goes as $t^{-1.5}$ instead of as t^{-2}. Individual cycles would only need to differ from the basic pulsation period by a few tenths of a minute to produce an observable effect on β. Can the theoreticians envisage this much disorder in the pulsation, perhaps through some interaction between pulsation and convection? Can any of us intuitively expect a fluid star to be more orderly than this in its pulsation?

A fourth possibility is another sort of cosmic noise: jumps in period which occur not at each cycle but separated by intervals of weeks, months or years during which the period is relatively stable. Abrupt changes are not attractive from a theoretical point of view, but this sort of behavior is strongly suggested by the many $O-C$ diagrams that look like several connected line segments, as Prof. Kukarkin reminded us in his review paper.

These four processes, which can be lumped together under the term random effects, are equally likely to cause negative as positive values of β. A fifth possibility to bear in mind is that the observations could be affected by some sort of error that has a systematic rather than a random effect on the calculated values of β.

Finally secular changes in the sizes of the stars are a sixth possibility for producing a $\beta \neq 0$, through the dependence of pulsation period on mean density.

The observational results are presented in summary form in Figure 1 as histograms showing the frequency distribution of the observed values of β, defined as $(1/P)(dP/dt)$ and expressed in cycles per million years. Look first at the lower figure. The mean epochs are 1893–98 (Bailey, 1902), 1931–35 (Martin, 1938) and 1963–67, the latter based on unpublished magnitudes from the Yale-Columbia Southern Station. A conspicuous feature of the distribution is its broad wings. Several of the rates are quite rapid, $+2.5$, -1.1, and -1.7 cycles per million years. Several more are between a half-cycle and one cycle in absolute value and there is one extreme case: V 104 is decreasing its period at the spectacular rate of 23 cycles per million years. Surely something besides evolution is going on here. Notice next the narrow core of the

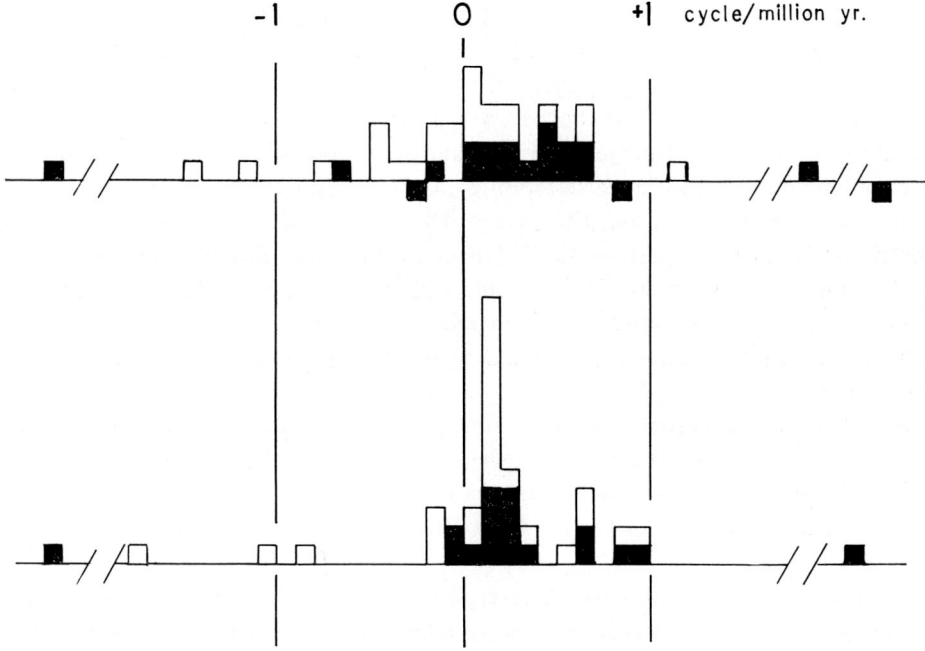

Fig. 1. Distribution of period changes observed in Omega Centauri. *Abscissa*: $\beta \equiv (1/P)/(dP/dt)$; *ordinate*: number of stars. Epochs for the lower histogram are 1893–98, 1931–35, 1963–67; epochs for the upper are 1931–35, 1947–48, 1963–67. The shaded portion is the part due to stars with periods longer than two-thirds of a day. The three boxes below the line in the upper figure are for stars that had not been observed in 1893–98.

distribution. About half of the stars seem to form a sub-group with slowly increasing periods, in the vicinity of two-tenths of a cycle per million years. Is evidence for evolution being heard above the noise, or can we account for the positive bias through systematic error and/or, by way of sampling error, through one or more of the random processes?

The median value of β is $+0.18$ cycle per million years. The mean is a less significant statistic because it is much more sensitive to the extreme values, which are very extreme indeed. Arbitrarily excluding V 104 and the two next most extreme on each side we find $\langle \beta \rangle = +0.24 \pm 0.05$ cycle. The dispersion of the 38 values around this mean is 0.32. Very little of the dispersion can be attributed to mean error of observation or fluctuation in epoch. The width of the central core may be due to these processes, but not the broad wings. The really large positive and negative values of β require a different explanation. Not evolution because the time scales are too short. If stars evolve so rapidly through the RR Lyrae stage they will be too uncommon in the population to contribute to our sample. We are almost certainly seeing some sort of average over the third and/or fourth of the random processes considered above, namely the accumulation of small cycle-to-cycle differences or larger less frequent abrupt changes.

Having decided that random processes are at work in producing some features of the observed distribution let us see whether, through sampling error, they can account for the positive bias. Prof. Kukarkin, in his review paper, asked for the application of modern mathematical statistics. I have applied only a simple Chi-square test, testing the hypothesis that the observed number of positive and negative values of β came about through chance selection of 43 cases from a population with equal numbers of positive and negative values. The observed numbers are 34 and 9, respectively. The probability is less than 5% that this much difference came about through such chance selection. A statistician would call this a significant deviation from the hypothesis tested and would ask us to reject the idea that positive and negative numbers are equally common in the parent population from which we have drawn our sample.

Statisticians and astronomers have different meanings for the word population. The result of the Chi-square test is not the same as denying that positive and negative period changes occur with equal frequency in a population consisting of all possible RR Lyr stars like these in Omega Cen. Are our numbers affected by systematic error?

Martin (1938) already applied whatever tests he could to his data to try to account for the positive bias by systematic observational error. His first epoch was Bailey's, the same one that I have used, but his second and third were his 1931 and 1935 observations considered separately. He redetermined the phase shifts between his 1931 and 1935 observations by a method that would eliminate the effect of any change in the magnitude system in the meantime. The result was a slight (but not significant) *increase* in the mean value of the rate of change of period. Then, because the first epoch was based on longer exposures, he considered whether the older observations might require a correction to account for distortion of the light curve. The required correction came out to be longer than the whole exposure time. We are now able to leave Bailey's observations out entirely if we still suspect that something might be wrong there. The upper portion of Figure 1 shows the result of calculating β from twentieth century observations only, using Yale-Columbia observations from 1947–48 as a mean point between the 1931–35 and the 1963–67 epochs. The dispersion is greater, as is to be expected, because the random processes have more influence on β when the time interval is shorter. The positive bias is still present, which shows that it had not been caused by some sort of systematic error in the 19th century observations. And yet something must be systematically wrong. The median is only +0.10 cycle per million years compared with +0.18. The median of Martin's values for these 43 stars was +0.30. Table I shows these along with other statistics of the three distributions. The β's increase, on the average, as we go from the latest back to the earliest selection of epochs. The differences seem a little too large to be merely statistical fluctuations. We are not going to suspect that stellar evolution was faster in the first third of the century than in the second! Apparently some unspecified systematic error has been at work to cause the differences. It can not, however, be the cause of the persistent positive bias. Table II shows the correction

TABLE I
Parameters of the distribution of β

Epochs	Median ($N=43$)	Mean ($N=38$)	Dispersion ($N=38$)	Obs. error (m.e.)
1931–35 1947–48 1963–67	+0.10	+0.11 ± 0.06	0.36	0.08
1893–98 1931–35 1963–67	+0.18	+0.24 ± 0.05	0.32	0.05
1893–98 1931 1935	+0.30	+0.33 ± 0.07	0.41	0.16

TABLE II
Hypothetical corrections, which, if applied to the midpoints of the exposures, would reduce the derived values of β by 0.1 cycle per million years

	1893–98	1931	1935	1947–48	1963–67
		-16^m		$+7^m$	-13^m
	-66^m	$+32^m$			-61^m
	-34^m	$+3^m$	-4^m		
ET-UT	$-0\overset{m}{.}1$	$+0\overset{m}{.}4$	$+0\overset{m}{.}4$	$+0\overset{m}{.}5$	$+0\overset{m}{.}6$
Exposure	30^m	10^m		65^m	18^m

to each epoch in each group of three that is required to reduce the resulting values of β by about 0.1 cycle per million years. Also listed are the values of ET-UT, which are negligible and have not been applied, and the lengths of typical exposures. Only the most artificial and unreasonable selection of 'corrections' would reduce the β's by the required amounts.

The observed values of β, then, are the results of random observational error, cosmic fluctuations in epoch and/or period, systematic error of unknown origin, and something else, something that can account for an average real increase in period at the rate of some two-tenths of a cycle per million years. The evidence for evolution in Omega Centauri is that secular increase in the sizes of the stars is a reasonable way to provide the increase.

Iben and Rood (1970) have shown that low-Z models with appropriate parameters of age, etc., arrive on the horizontal branch to the blue of the instability strip and cross it evolving to the red with periods increasing at the rate of a few tenths of a cycle per million years, which is in agreement with the observations in Omega Cen provided that we attribute the negative observed values to larger, non-evolutionary fluctuations superimposed on the secular changes. If their theory is correct there is a range in mass, with the lower-mass stars crossing the strip more rapidly and at higher luminosities. The observations do not show a correlation of β with magnitude but perhaps this is only because magnitudes of variable stars in a crowded field are not precise enough. The mass range should also produce a correlation of β with period since the less massive, more luminous stars are less dense as they cross the strip and therefore of longer period. This correlation has been observed (Martin, 1938; Belserene, 1964); the shaded portion of the histograms in Figure 1 is the part due to stars with periods longer than two-thirds of a day. This portion does indeed favor increasing periods more strongly than the rest. A roundabout way of looking for a relation between β and luminosity is with the help of a period-amplitude relation such as Figure 2, where the kind and size of symbol shows the sign and size of β. The positive values of β and particularly the larger ones tend to occur not only at the longer periods but especially where the period is long for the amplitude. We can interpret this as the correlation of β with magnitude if we simply look at the diagram from another angle. Tilt the page counterclockwise until the lines of constant period

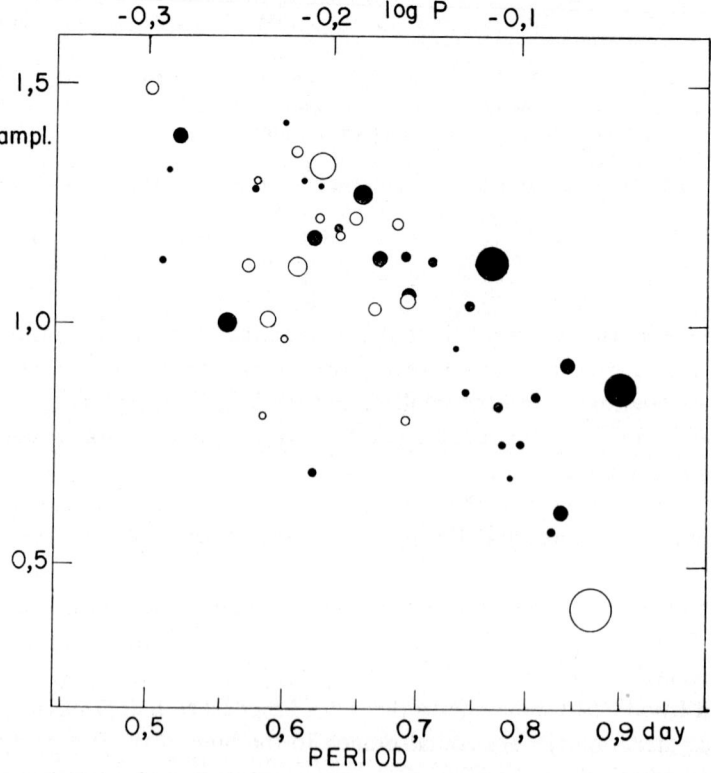

Fig. 2. Period-amplitude relation for 46 regular RR$_{ab}$ variables in Omega Centauri. *Filled circles*: increasing periods; *open circles*: decreasing periods; larger dots mean faster rates. The rates are from the epochs: 1931–35, 1947–48, 1963–67.

are more nearly horizontal than vertical. This is rather like the way the lines of constant density cross the horizontal branch. The lines of constant amplitude now slant steeply up to the right, and the line of zero amplitude has become the red edge of the instability strip. Viewed in this way the period-amplitude plane has become a somewhat distorted version of a piece of the color-magnitude plane. The stars have entered the RR_{ab} region on their way to the right from a blue horizontal branch and the more luminous ones are going fastest. There are many exceptions in Figure 2, but we account for most of those by the superposition of random on secular processes, especially at the shorter periods near the transition from RR_c.

Such is the evidence for evolution in Omega Cen: present but not very strong. This is not to say that evolution has been at work whenever there is curvature in an $O-C$ diagram. In fact, when the curvature is obvious the rate is usually too fast to be secular. To say that evolution has been detected in Omega Cen is not the same as claiming that it accounts for all period changes observed in this and other clusters or even that it can ever be detected in other clusters. Omega Centauri is very unusual, both variable-rich and variable-poor, rich in number of variable stars but poor when this number is normalized to account for the total population of the cluster (Belserene, 1956; Kukarkin, this colloquium). Only a variable-rich cluster will have a large enough sample of stars to show a trend in the presence of noise. Even then the rate of evolution will be orders of magnitude slower than the noise unless the stars pass through the RR Lyrae phase rather rapidly. If they do, the cluster is necessarily poor in variables in proportion to the total number of stars. Omega Centauri is a very favorable case.

References

Bailey, S. I.: 1902, *Ann. Harvard Coll. Obs.* **38**.
Belserene, E. P.: 1956, *Contr. Rutherfurd Obs.* No. 33, 13.
Belserene, E. P.: 1964, *Astron. J.* **69**, 475.
Iben, I., Jr. and Rood, R. T.: 1970, *Astrophys. J.* **161**, 587.
Martin, W. C.: 1938, *Ann. Sterrew. Leiden* **17**, Part 2.
Sterne, T. E.: 1934, *Circ. Harvard Coll. Obs.* No. 386.

DISCUSSION

Cox: Do these studies by you and the others include the RR_c variables? Do we have evidence of any *c*-type star changing to an *ab*-type or vice-versa?

Belserene: There were no RR_c's in this sample. They have larger fluctuations in period and are difficult to follow. We haven't been looking at the cluster long enough to expect to catch a change in type.

Jones (to Cox): AC And is the only RR Lyrae in the field with a period ratio of roughly 4:3 which indicates that the fundamental and first overtone are simultaneously excited. It has no photoelectric light curve. I know of no RR Lyrae which has changed its Bailey type.

Schwarzschild: It seems to me that all the papers we heard this morning agree on one point: Random variations of the period of RR Lyr variables are common, if not the rule – a real challenge to theory. Since these random changes appear percentage-wise to be very small they cannot be observed in any other quantity, which makes the determination of the character of the random period changes extraordinarily important.

Wesselink: Do stars with Blazhko effect behave differently from the others with respect to period behaviour?

Belserene: The stars known to have a Blazhko effect were not on this program. Unrecognized Blazhko-effect stars probably contribute to the high-dispersion component in the distribution of the period changes.

Aizenman: Is it possible that high order harmonics which may be generated in the atmosphere might interfere with the fundamental mode in a manner that gives rise to a time variation in the maximum of the light curve?

Cox: In all accurate non-linear calculations that I know about, the star pulsates in only a pure mode. An atmospheric phenomenon probably cannot excite a deepseated variation such as these pulsations.

SHORT PERIOD VARIABLES IN THE MAGELLANIC CLOUD CLUSTER NGC 1466

M. V. NORRIS
Dunsink Observatory

1. Introduction

NGC 1466 ($\alpha_{1950} = 3^h44^m.6$, $\delta_{1950} = -71°45'$) is a globular cluster which appears to be situated between the two Magellanic Clouds. Previous estimates (Gascoigne, 1966) put it at roughly the same distance from us as the LMC, so it is regarded as a member of the Cloud system. It is globular in appearance, and its colour-magnitude diagram confirms this classification. It has a fairly well-developed horizontal branch, and was found by Wesselink (1970) to be quite rich in variables. The metallicity index, Q, (van den Bergh, 1967) has a value of -0.36 for NGC 1466 (Andrews and Lloyd Evans, 1971). This would rank it with M5 and NGC 6171 as a cluster of intermediate metal content. This comparison is consistent with the value of ΔV for the cluster, which, at $2^m.6$, is representative of the ΔV values of globular clusters of intermediate metal abundance in the Galaxy.

2. Observations

The observational material consists of three sets of plates, taken with different photometric systems:

(i) 21 plates taken in October–December 1969 on the 188-cm telescope of the Radcliffe Observatory. 12 of these were V plates and 9 were B plates.

(ii) 11 plates taken by Wesselink in October–November 1952 on the Radcliffe telescope, using 103a-O emulsion with no filter.

(iii) 4 B plates taken by John Graham in November 1969 with the 152-cm telescope at Cerro Tololo.

The three sets of plates were measured on an Askania iris photometer. The faint photoelectric sequence of Gascoigne (1969) was used to perform the reductions. The effects of varying background density were partially eliminated by means of a method outlined by Weaver (1962).

Because of the quasi-systematic way in which colour varies with magnitude in Gascoigne's sequence, it was not possible to use this sequence to determine colour terms for the Radcliffe B and V plates. Instead, Butler's OL sequence in LMC II (Butler, 1971) was used for this purpose. B and V plates of this sequence had been taken with the Radcliffe telescope in 1969, so that observations of the OL sequence, in the same photometric system as the observations of NGC 1466 were available.

The unfiltered Wesselink plates were reduced by means of a secondary sequence, determined from the reduced measurements of the B and V plates. The colour term, which was quite large and negative, was also derived by means of this secondary sequence. The Cerro Tololo plates were measured solely to provide more points on the light curves of the variables. Therefore, only confirmed variables and the stars of Gascoigne's sequence were measured on these plates.

Periods of variables were determined using the B, V and Wesselink sets of plates individually. Periods were adopted only if they agreed between at least two sets of plates; in fact, most of the periods found agreed between the three sets. Periods were found for 9 *ab*-type RR Lyrae variables, and one *c*-type was tentatively confirmed. Of these 10 variables, periods had been found for 6 by Wesselink (1971). The periods found for these 6 in the present work agree, to within the limits of accuracy, with Wesselink's values. The period of No. 98, the *c*-type variable, can be confirmed as $0\overset{d}{.}353$ if it is assumed that the amplitude varies from cycle to cycle.

The number of plates in either colour was not sufficient to construct separate light curves, so it was decided to combine the B, V, Wesselink and Cerro Tololo plates to construct one light curve, in B, for each variable. This was done by assuming a mean variation of colour with phase, derived from observations of RR Lyrae stars in NGC 6171 by Dickens (1970), and a mean colour derived from rough light curves in B and V. In this way, V magnitudes could be converted into B magnitudes at the same phase:

$$B_\phi = V_\phi + (B-V)_\phi + (\langle B \rangle - \langle V \rangle), \qquad (1)$$

where $(B-V)_\phi$ is a smooth normalised function derived from the mean colour curves of the *ab*-type variables in NGC 6171. For the *c*-type variable in NGC 1466, $(B-V)_\phi$ was derived from the *c*-type colour curves in NGC 6171.

The colour derived from the light curves is given by $\langle B \rangle - \langle V \rangle$. This is not the same as the mean of the colour intensity curve, $\langle B-V \rangle$. In fact, on examining Dickens' data, one finds a systematic difference between the two colours, such that $\langle B-V \rangle$ is always redder than $\langle B \rangle - \langle V \rangle$. The mean difference is

$$\langle B-V \rangle - (\langle B \rangle - \langle V \rangle) = +0\overset{m}{.}03. \qquad (2)$$

NGC 1466 is free of any absorption by either Magellanic Cloud, and is subject only to a small amount of Galactic absorption. Extrapolating from the data of Butler (1971) in the wing of the LMC, we assign a reddening of $+0\overset{m}{.}02$ to the cluster. Thus, the $\langle B \rangle - \langle V \rangle$ values are transformed to the unreddened mean colours, $\langle B-V \rangle_0$, by adding $0\overset{m}{.}03 - 0\overset{m}{.}02 = 0\overset{m}{.}01$.

3. Results

All the light curves, except that of No. 98 resemble Bailey type-*ab* RR Lyrae variable light curves, in that they have a sharp rise to maximum light, and a slow descent. No. 109 with the shortest period of the *ab*-types, has a rising branch which is not extremely steep, while No. 116 is an extreme example of an *a*-type variable, with a pronounced dip just before the steep rise to maximum (Figure 1).

The mean period of the *ab*-type variables is $0\overset{d}{.}533$, indicating that NGC 1466 is a cluster of Oosterhoff Group I (Oosterhoff, 1939). So far, no clusters of Oosterhoff

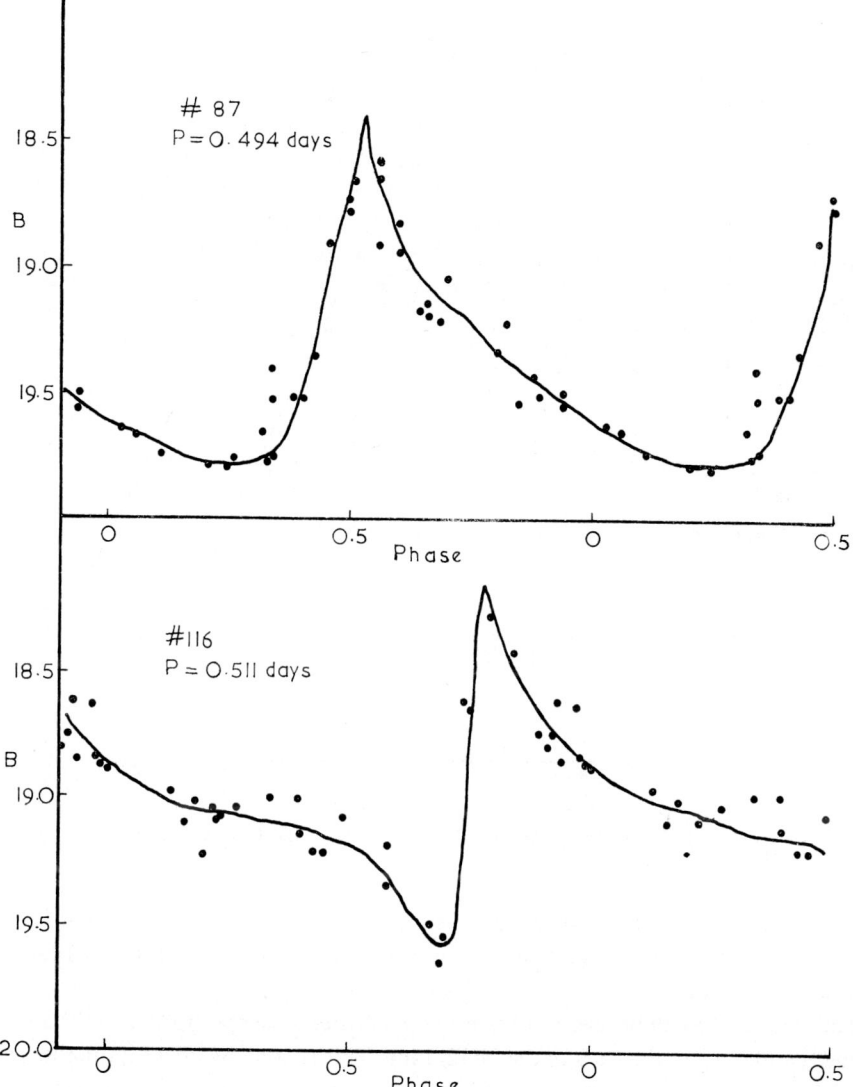

Fig. 1. Light curves of two variables in NGC 1466.

group II, with longer mean *ab*-type periods, have been found in the Magellanic Clouds, suggesting that this system is not as evolved as the Galaxy.

Preston (1959, 1961a) has found that the period-amplitude diagram for a group of RR Lyrae variables depends on the value of the line strength index, ΔS, for the group. The period-amplitude relation for variables in NGC 1466 (Figure 2) lies roughly between those of variables in M5 ($\Delta S=3$) and variables in M3 ($\Delta S=6$), i.e. it corresponds roughly to a ΔS-value of 4.5. Using values of $[M/H]$ found by Preston (1961b) for certain field RR Lyrae variables, and values of ΔS of the same stars (Preston, 1959),

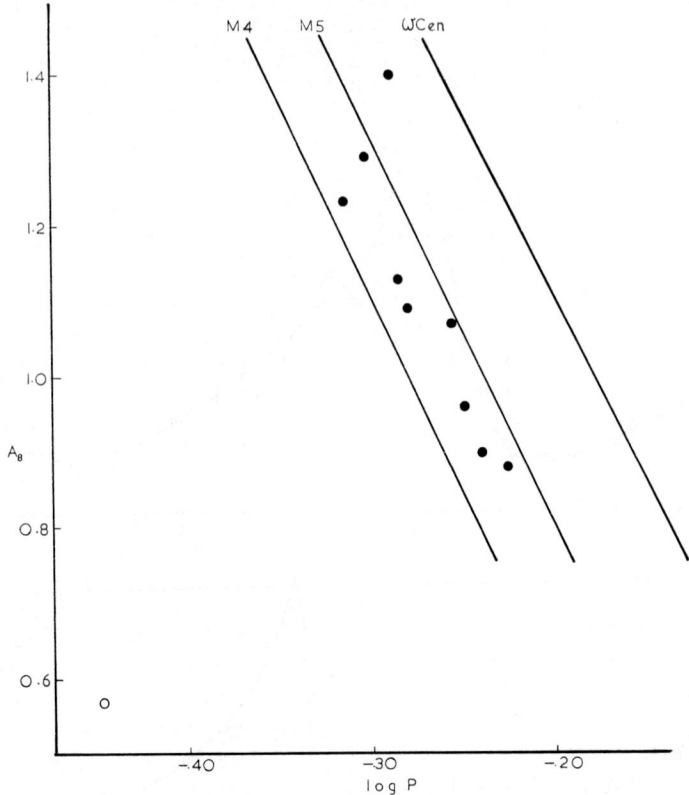

Fig. 2. Period-amplitude diagram of variables in NGC 1466.

these two quantities can be correlated. Using such a relation, a ΔS-value of 4.5 corresponds to a heavy element underabundance by a factor of 6 or 7 in NGC 1466 with respect to the Sun.

The value of ΔS can be used to obtain a temperature scale appropriate to the amount of line-blanketing in the stars in the NGC 1466. This is done by interpolating in Table II of Dickens (1970) between ΔS-values of 3 and 6, and linearizing the scale in the $\log T_e - (B-V)_0$ plane. Using this scale, together with the $\log P - (B-V)_0$ array for the variables (Figure 3) it is possible to derive a mass-luminosity relation from the $P\sqrt{\varrho}$ = constant analogue of van Albada and Baker (1971):

$$\log P_f = -1.772 - 0.68 \log M + 0.84 \log L + 3.48 \log \frac{6500}{T_e}, \qquad (3)$$

where M and L are in solar units. This procedure yields a relation

$$\log \frac{M}{L} = -1.89 - 0.094 M_{bol}. \qquad (4)$$

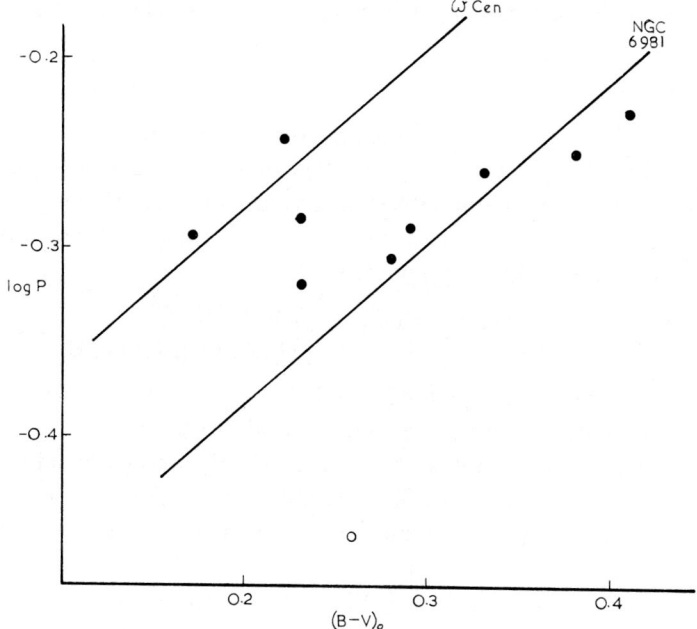

Fig. 3. Period-colour diagram of variables in NGC 1466.

When the value of $M_{bol} = +0\overset{m}{.}466$ (see below) is inserted, a mass-luminosity ratio is obtained:

$$\log \frac{M}{L} = -1.93 \pm 0.09. \tag{5}$$

The absolute magnitude of the variables is derived from Equation (16) of Iben (1971), which is obtained from the results of linear pulsation theory models:

$$M_{bol} = (1.98\,Y - 2.5 \log P_I - 1.07)/(0.96 - 1.1\,Y), \tag{6}$$

where P_I is the fundamental period at the intersection of the fundamental and first harmonic 'blue edges' in the luminosity-colour diagram. This is equivalent to the shortest fundamental, or *ab*-type, period in a cluster, which in the case of NGC 1466, is given by

$$\log P_I = -0.320. \tag{7}$$

Inserting this value in Equation (6) gives

$$M_{bol} = (1.98\,Y - 0.28)/(0.96 - 1.1\,Y). \tag{8}$$

Equation (8), when combined with Equation (12) (see below), yields the following value for the absolute magnitude of the variables:

$$M_{bol} = +0\overset{m}{.}46. \tag{9}$$

combining this with the mean apparent magnitude of the variables yields an uncorrected distance modulus for the cluster of

$$(V - M_{bol}) = 18^m.70 \pm 0^m.15 \tag{10}$$

the error being due to the scatter in the apparent magnitudes of the variables. This value for the distance modulus agrees with that found by Gascoigne (1966).

The helium content of the variables is derived from the relation governing the temperature at the blue edge of the instability strip, obtained empirically by Iben (1971) from the results of his linear models:

$$\log L = 1.84 - 5.08\Delta Y - (12.3 - 61\Delta Y)\Delta \log T_e + \\ + [(0.9 - 0.4\Delta Y) - (16 - 33\Delta Y)\Delta \log T_e] \log M, \tag{11}$$

where $\Delta Y = Y - 0.3$ and $\Delta \log T_e = \log T_{eBE} - 3.87$.

T_{eBE} = effective temperature at blue edge of instability strip.

Using the temperature scale described above, and the observational evidence that the unreddened mean colour of the bluest variable in the cluster is $+0^m.17$, together with the previously derived mass-luminosity relation (Equation (4)), Equation (11) reduces to

$$0 = -0.110 + 5.386\Delta Y + 0.351 M_{bol} + 0.116\Delta Y \cdot M_{bol}. \tag{12}$$

When combined with Equation (8), this yields

$$Y = 0.29, \tag{13}$$

i.e., a normal Helium content for a Population II object in the Magellanic Clouds.

Acknowledgements

The author is indebted to Prof. P. A. Wayman for his encouragement and invaluable help during the present work. Thanks are also due to the Dublin Institute for Advanced Studies for their financial support.

References

Albada, J. D. van and Baker, N. H.: 1971, *Astrophys. J.* **169**, 311.
Andrews, P. J. and Lloyd Evans, T.: 1971 in A. B. Muller (ed.), *The Magellanic Clouds*, Reidel Co., Dordrecht, p. 88.
Bergh, S. van den: 1967, *Astron. J.* **72**, 70.
Butler, C. J.: 1971, Thesis, Trinity College, Dublin.
Dickens, R. J.: 1970, *Astrophys. J. Suppl.* **22**, 249.
Gascoigne, S. C. B.: 1966, *Monthly Notices Roy. Astron. Soc.* **134**, 59.
Gascoigne, S. C. B.: 1971, Private communication.
Iben, I.: 1971, *Publ. Astron. Soc. Pacific* **83**, 697.
Oosterhoff, P. Th.: 1939, *Observatory* **62**, 104.
Preston, G. W.: 1959, *Astrophys. J.* **130**, 507.
Preston, G. W.: 1961a, *Astrophys. J.* **134**, 651.
Preston, G. W.: 1961b, *Astrophys. J.* **134**, 633.
Weaver, H.: 1962, *Handbuch der Physik* **54**, 130.
Wesselink, A. J.: 1970, Private communication.
Wesselink, A. J.: 1971, *Monthly Notices Roy. Astron. Soc.* **152**, 159.

DISCUSSION

Graham: (1) Is not NGC 1466 slightly nearer to the LMC than the SMC? (2) Did you find any faint RR Lyraes in the general field outside the cluster?

Norris: In terms of angular separation, NGC 1466 is closer to the Large than to the Small Cloud. This, combined with the fact that the cluster appears to be at the same distance from us as the LMC, would place NGC 1466 linearly closer to the LMC than to the SMC. (2) The Radcliffe plates have been blinked almost to the edges, i.e., to about 70′ from the centre of NGC 1466. No new variables were found in this region, which at the distance of the cluster, has a radius of about 300 pc.

THE RR LYRAE STARS IN THE MAGELLANIC CLOUDS

J. A. GRAHAM

*Cerro Tololo Inter-American Observatory**

Abstract. The characteristics of the RR Lyrae stars in the Magellanic Clouds are discussed. The existence of numerous RR Lyrae variables in the Large Magellanic Cloud (LMC) cluster NGC 1835 is noted. The variables indicate that this cluster is of Oosterhoff type I. No cluster of Oosterhoff type II has yet been recognized in the Clouds. Some new results of a study of the RRLyrae variables in an LMC field $1° \times 1°.3$ are discussed. Periods have been determined for 50 out of 72 probable RR lyraes in this field. RR Lyrae *ab* stars with periods less than $0^d.46$ are absent. The time averaged $\overline{\langle B \rangle}$ and $\overline{\langle V \rangle}$ are $19^m.56$ and $19^m.20$ with surprisingly small dispersion. First impressions of results for a similar field in the Small Magellanic Cloud (SMC) suggest that the old stellar population of the SMC may have a slightly lower metal abundance than that of the LMC. The best available distance moduli for the Magellanic Clouds indicate a mean absolute visual magnitude $M_{\langle \bar{v} \rangle}$ of $+0^m.5 \pm 0.2$ for the RR Lyrae stars in these systems.

1. Introduction

When we come to study the older populations of the Magellanic Clouds, we are no longer very concerned with the well-known spectacular features which dominate most photographs of these stellar systems. We must look past the brilliant associations with their blue supergiants and H II regions, past the Cepheid variables and the numerous open star clusters until we see in each Cloud only the faint amorphous substrata which are made up of stars and planetary nebulae with ages 10^9 yr or more. Relieving the general uniformity, several old globular clusters, similar to those of our own Galaxy are seen, scattered among the two Clouds. Except for the occasional nova every star in this old population is fainter than 15th magnitude. However, despite their rather dull appearance, these faint substrata of low mass stars are indeed the structural backbones of the Magellanic Clouds. They probably contain a significant fraction of the mass of each system and thereby are important contributors to the internal dynamics of the Clouds. It is also important to realise that the study of these old populations may give information about the history of nuclear synthesis in the Clouds back to the earliest stages of their evolution.

Detailed investigation of the faint Population II stars of the Magellanic Clouds will have to wait for the new 4-m reflectors to begin work in the Southern Hemisphere. However, quite a bit of work can be done now. In a recent review article, Westerlund (1970) has discussed much of our present knowledge of the Cloud populations. The planetary nebulae have been studied in considerable detail by Feast (1968), Webster (1969), and Smith and Weedman (1972). Novae are being detected at the rate of 1 or 2 a year at Cerro Tololo Inter-American Observatory by Graham and Araya (1971). Gascoigne's (1966) investigation of the color-magnitude diagrams of the Cloud globular clusters has recently been extended by Walker (1970, 1971, 1972),

---

* Operated by the Association of Universities for Research in Astronomy, Inc., under contract with the National Science Foundation.

who has made use of electronographic techniques to reach stars as faint as blue magnitude 23.5. In the present article, the characteristics of the RR Lyrae-type variables are described. These stars are perhaps the easiest of the Population II objects to discover because of the short period and substantial amplitude of their light variation. In the context of this colloquium, I think that it is important to stress that not only can the RR Lyraes tell us about the Magellanic Clouds, but that also the Magellanic Clouds can tell us quite a lot about RR Lyrae stars. Among the globular clusters and related systems that we are concerned with at this conference, the Magellanic Clouds alone offer the opportunity to compare, at effectively the same distance and often on the same photographic plate, an abundance of the youngest Population I and the oldest Population II stars.

RR Lyrae stars were first detected in the Magellanic Clouds by Thackeray (1951, 1958) who found three in the Small Magellanic Cloud (SMC) cluster NGC 121 and three in the surrounding field outside the probable boundary of the cluster. Thackeray also discovered RR Lyraes in the Large Magellanic Cloud (LMC) globular clusters NGC 1978 (Thackeray, 1951), NGC 1466 (Thackeray and Wesselink, 1953), and later in NGC 2257 (Alexander, 1960). Further photometry of these variables has been carried out by Alexander (1960), Tifft (1963), Gascoigne (1966), Wesselink (1971), and Norris (1972).

2. Cerro Tololo Investigations

The recent availability of large reflecting telescopes with wide photographic fields has made much easier the discovery of the RR Lyrae variables in the general field of the Magellanic Clouds. With the CTIO 1.5-m reflector, stars of blue magnitude 20.5 can be recorded with a 30-min exposure over a field of $1°\!.5$ on 103a-O plates. The $f/7.5$ scale of the telescope is $18.1''$ mm^{-1}. With 20 cm × 25 cm plates an effective field of $1° \times 1°\!.3$ is obtained. Vignetting appears to be less than $0^m\!.1$ over this area.

There are four principal fields for which series of plates are available. One field is centered on the LMC cluster NGC 1783. This region was chosen because it contains the photoelectric sequence determined by Gascoigne (1962). A second field, in the vicinity of the LMC bar, is between the globular cluster NGC 1835 and the photoelectric sequence of Tifft and Snell (1971). In the SMC, many plates have now been taken of a field centered on the globular cluster NGC 121. Tifft (1963) has published a sequence going to $B=20^m\!.8$ close to this cluster. Finally, some plates have been taken of a region closer to the SMC bar. These are centered between NGC 361 and NGC 362. 103a-O plates were initially used to obtain the maximum number of plates and the greatest time resolution. However, since it became clear that precise blue magnitudes are very important, longer exposures with 103a-O + GG 13 filter have been taken. For each field two or three yellow sensitive plates (103a-D + GG 14) have been obtained. These are useful for the determination of mean V magnitudes for the RR Lyrae variables, as well as for the identification of very red stars in the field.

The variables are discovered by comparing pairs of plates, taken under similar conditions a week or more apart. For each field it is planned eventually to compare

10 such pairs. Following the discovery, the variables and standard sequence stars are measured on each plate with an iris-diaphragm photometer. The transformation curve relating iris-diaphragm reading and magnitude is determined by eye estimate only.

3. RR Lyrae Stars in the Magellanic Cloud Globular Clusters

To the body of work already published, I am able to add some new data for the variables in an additional cluster, NGC 1835. Sra. Maria Teresa Ruiz, a former student at the University of Chile, has helped me considerably in this part of the investigation. NGC 1835 is a very compact cluster quite close to the LMC bar. Because of crowding both in the cluster and in the surrounding dense star field, we have only been able to use plates taken under the best seeing conditions. Although much of the cluster is unresolved, it is clear that it is very rich in variables. Forty-nine stars were suspected of variation from the blink comparisons. Subsequent iris-diaphragm photometer measurements have shown that for twenty-one of the forty-nine, the variation is certainly real. Periods have been found for ten stars. Nine of these are RR Lyrae ab-type stars, and one a c-type. Among the eleven cases remaining, there is one long period variable, one possible Cepheid and a few more RR c-type variables. The standard magnitude sequence for this cluster is calibrated in two ways: first by a photographic transfer from the Gascoigne (1962) NGC 1783 sequence and second by direct measurement of the nearby Tifft-Snell sequence on the same plate. Agreement is found within $0^m.1$ for the stars calibrated in both ways.

In Table I, I summarise the presently available data for the Cloud cluster RR Lyraes. The arithmetic mean of the ab-type variable star periods, the arithmetic means of the apparent blue median magnitude, of the time-averaged blue magnitudes

TABLE I
RR Lyraes in Magellanic Cloud globular clusters

Cluster		No. of variables with periods	$\bar{P}_{ab}$	$\bar{B}_{med}$	$\langle B \rangle$	V_{RG}	Ref.
SMC	121	3	$0^d.560$	$19^m.55$	$19^m.7$	$16^m.7$	(1)
	1466	10	0.549		19.23	16.2	(2) (3) (4)
LMC	1835	10	0.572	19.25	19.40	16.0	(5)
	2257	6	0.558	19.19	19.31	16.1	(2) (6)
			0.559	19.22	19.31	16.1	

References:
(1) Tifft (1963).
(2) Gascoigne (1966).
(3) Wesselink (1971).
(4) Norris (1972).
(5) Graham and Ruiz (unpublished).
(6) Alexander (1960).

and of the apparent visual magnitude at the red giant branch tip are given where possible for each cluster. One notes immediately that all the above clusters are of Oosterhoff type I on the basis of the variable star data. Although the numbers are still small, it seems possible that Oosterhoff type II clusters may turn out to be rare or absent in the Magellanic Clouds, indicative perhaps of the absence of an extremely metal poor population. A similar suggestion comes from the work of Andrews and Lloyd Evans (1971) who have found spectra indicating only moderate metal deficiency for a sample of thirteen Magellanic Cloud clusters, In Table I, it should also be noted, a comparison between the three LMC globulars and the one SMC globular suggests that the SMC has an apparent distance modulus $0\overset{m}{.}4 \pm 0.2$ greater than that of the LMC.

4. The Field RR Lyrae Stars in the Magellanic Clouds

For the four fields mentioned in Section 2, only the analysis for the NGC 1783 field has progressed sufficiently so that good data are now available for the variables in this field. This region of the LMC is of special interest as it contains not only the intermediate age cluster NGC 1783 but also a strong Population I component in the form of numerous Cepheids, blue supergiants and prominent H II regions. Clearly, in this part of the LMC we are viewing stars with a very wide range in age. To detect the RR Lyrae stars, ten pairs of plates were blinked and 125 stars marked as suspected variables. Of these, only one, an eclipsing variable, is a possible member of NGC 1783 itself (Graham, 1970). The iris-diaphragm measurements indicate that about 50 are Cepheid, red, or eclipsing variables, 72 are probable RR Lyrae stars. Periods have been determined for 51 of the 72 suspected RR Lyraes. Only one of the 51 is a c-type variable, although a few of the 21 difficult cases may also be RR Lyraes of this type. There seems little doubt, however, that the c-type variables are relatively rare compared to the ab-type.

Table II summarises the characteristics of the NGC 1783 sample. Note the absence of the RR_{ab} stars with periods less than $0\overset{d}{.}46$, which form about 25% of the RR_{ab} population in the vicinity of the Sun.

TABLE II

RR Lyrae stars in the NGC 1783 field in the LMC

Number of probable RR Lyraes	72
Number of RR_{ab} stars with periods	50
Number of RR_c stars with periods	1
Most frequent period between $0\overset{d}{.}50$ and $0\overset{d}{.}55$.	
RR_{ab} stars with $P < 0\overset{d}{.}46$ absent.	
Period-amplitude diagram similar to that of Messier 3 but with more dispersion.	
$\bar{B}_{median} = 19\overset{m}{.}35$	
$\overline{\langle B \rangle} = 19\overset{m}{.}56$	
$\overline{\langle V \rangle} = 19\overset{m}{.}20$	

The mean magnitudes given in Table II are a tenth of a magnitude or so fainter than the cluster values in Table I. A similar result was found by Tifft in a comparison between the RR Lyraes in the SMC cluster NGC 121 and those in the general field. It seems likely that this effect is largely due to background enhancement arising from faint, unresolved stars and from light scattered by the bright cluster center, and that the values derived from field star observations are the more accurate ones.

A surprising result of the present investigation is the small dispersion in the time averaged blue magnitude, $\langle B \rangle$, among the LMC RR Lyraes. In Figure 1, two histograms are presented comparing the dispersion of the NGC 1783 field variables and of the Messier 3 (M3) variables (Sandage, 1959). It should be noted that the mean magnitudes of the M3 variables are intensity means, while those presently available for the LMC variables are time-averaged magnitudes. However, this should not affect the present comparison. The dispersion among the Cloud RR Lyraes appears to be only slightly greater than that found among the M3 stars. Besides intrinsic dispersion, the dispersion among the Cloud sample must include such factors as interstellar absorption within the Cloud, scatter in distance along the line of sight (at the LMC distance

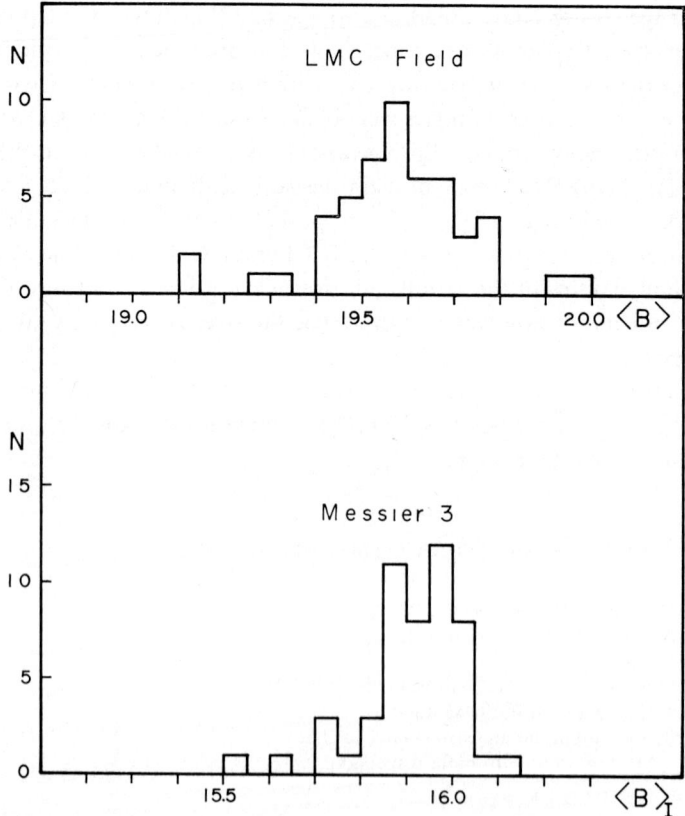

Fig. 1. Frequency distribution of the apparent blue magnitudes of the RR Lyrae stars in the LMC field near NGC 1783 and of the RR Lyrae stars in the Galactic globular cluster Messier 3.

±1 kpc = ±0^{m}04) as well as photographic plate errors. The small dispersion found indicates that (1) the intrinsic dispersion of the RR Lyrae absolute magnitudes is small, not exceeding the dispersion found among the M3 variables by more than 0^{m}1 and (2) probably the Cloud RR Lyraes are concentrated in a disk system rather than spread through an extended halo system. The Christy pulsation theory (summarised by Christy (1966)) predicts that in the Galaxy the observed range in metal abundance will produce a range in absolute magnitude for the field RR Lyraes of about 0^{m}4. The low dispersion found for the absolute magnitudes of the LMC variables suggests that if this theory is correct, the range of metal abundance may be quite small among the old Cloud population. Some support for this suggestion comes from the observed absence of the short period RR_{ab} variables with $P < 0^{\mathrm{d}}46$ days and, at the other extreme, by the possible absence of Oosterhoff type II globular clusters with weak-lined integrated spectra.

It is possible to obtain first impressions of the characteristics of the RR Lyraes in the SMC using some measurements of the available plate material for the NGC 121 field. More plates are still required before a detailed analysis can be undertaken. Table III summarises the SMC situation at present. If the first impressions are correct, then from comparing Tables II and III, there do seem to be some significant differences between the RR Lyraes in the two Magellanic Clouds.

TABLE III

First impressions of the RR Lyrae stars in the NGC 121 field in the SMC

Number of Probable RR Lyraes	60
Number of RR_{ab} stars with periods	16
Number of RR_c stars with periods	4
Most frequent period between $0^{\mathrm{d}}55$ and $0^{\mathrm{d}}60$	
RR_{ab} stars with $P < 0^{\mathrm{d}}50$ rare or absent.	
$\bar{B}_{\mathrm{median}} \approx 19^{\mathrm{m}}8$	
$\overline{\langle B \rangle} \approx 20^{\mathrm{m}}0$	
$\overline{\langle V \rangle} \approx 19^{\mathrm{m}}6$	

In the context of the Christy theory, the transition point between the fundamental and first overtone modes of pulsation seems to be shifted relative to the LMC by about $0^{\mathrm{d}}05$ towards a longer period in the case of the SMC. This may indicate that the old population in the SMC is slightly metal poor, on the average, compared with that of the LMC. The values for the mean magnitudes are still uncertain, and, considering the preliminary state of the analysis, the agreement with the earlier work of Tifft (1963) is quite reasonable.

5. The Absolute Magnitudes of the Magellanic Cloud RR Lyrae Stars

The distances of the Magellanic Clouds are now sufficiently well known for it to be possible to make a good estimate of the absolute magnitude of their RR Lyrae vari-

Graham: It is hard to believe that the errors involving the Magellanic Cloud distance moduli can be large enough for this to be true.

Buscombe: Are the blue globular clusters of the LMC related to old or young populations?

Evans: Andrews and I have measured radial velocities of 14 of these clusters and find the velocity dispersion is not different from that of the entire Population I.

Buscombe: Are the 'blue globulars' kinematically related to Population II, or merely very rich open clusters?

Graham: Holland Ford in his thesis found, I believe, that they had Population I characteristics.

Cooper: I believe Soviet work has determined some absolute magnitudes for RR Lyrae stars to be considerably closer to 0.0 than 0.5. Can you comment on this?

Graham: I am not aware of these results.

PART III

SLOW VARIABLES IN POPULATION II SYSTEMS

OBSERVATIONAL ASPECTS OF SLOW VARIABLES IN GLOBULAR CLUSTERS

M. W. FEAST

Radcliffe Observatory, Pretoria, South Africa

Abstract. There are up to 14 known Mira variables in seven globular clusters, though several have not yet been confirmed as radial velocity members. The periods of only 5 are known, all near 200 days. The clusters seem to form a compact group of relatively metal rich clusters. In 3 or 4 cases spectroscopy shows that the giant branches of these clusters penetrate into the M types. The Mira-containing clusters also contain red variables of shorter period and smaller amplitude which are generally also M type stars. Stars apparently evolve to the red of the giant tip as variables of increasing amplitude and period. Effects of TiO blanketing on the $(B-V)$ colours may be anticipated in these clusters.

Besides variables at the red giant tip the metal poor globular cluster ω Cen contains variables with strong TiO bands. Photometry, including recent J, H, K, L photometry by Glass shows that these stars are very cool objects. They indicate an extension of the giant branch considerably cooler than previously considered for metal poor clusters.

V1, NGC 121 in the Small Magellanic Cloud has a spectrum indicative of an SRd variable. It is not yet clear whether galactic stars similar to this star exist or not.

1. Introduction

The title of this review covers, in effect, all kinds of variables except RR Lyrae stars. However in order to give some coherence to the paper, the discussion will be almost entirely limited to the red giant variables; that is to variables in the general region of the tip of the red giant branch in the HR diagram. These stars are of considerable importance in helping to understand what is going on at, and beyond, the red giant tip. As yet there is rather little detailed theoretical work on these stars, but with the increasing amount of observational work being carried out on them we may hope that more theorists will be tempted into this field.

2. Mira Variables in Clusters

Let us first consider the membership of Mira variables in clusters. For our purposes we define Mira variables as red giants with long periods (generally greater than 100 days), large amplitudes (Δm greater than about $2^{m}5$) and Me spectra (that is spectra showing TiO bands and emission lines, generally Balmer lines, at least at some phases). It is quite instructive as a start to look at a list (Table I) of Mira variables which lie in the direction of globular clusters and which have been shown to be non-members of the clusters. Generally, of course, membership is decided on the basis of radial velocity work, which is usually unambiguous since the clusters and, to a lesser extent, the Mira variables have large peculiar motions and chance agreement of velocity is unlikely. It is worth noticing that the Mira variables in Table I scatter over quite a large range in periods (177–334 days). When we look at Miras that are members of clusters we find that they cover a much more restricted range of periods.

TABLE I
Field Miras in the direction of globular clusters

Cluster		Variable	Period (days)	ΔV (km s^{-1})	Reference
NGC 6656	M22	V14	200	201	Astrophys. J. **110**, 105.
4590	M68	FI Hya	324	214	Publ. Astron. Soc. Pacific **59**, 143.
4833		RZ Mus	334	252	Observatory **86**, 120.
6397		V1	315	30	Observatory **86**, 120.
5139	ω Cen	V2	242	281	Observatory **85**, 16.
6171		V720 Oph	332	>100	Dickens (private communication).
6093	M80	S Sco	177	67	
6093	M80	R Sco	222	19[a]	

ΔV is the difference between the cluster radial velocity and the radial velocity of the variable.
[a] This value of ΔV is not large enough to rule out membership but the star is far out from the cluster making membership unlikely. If it were a member it would be an extremely bright object ($M_v \sim -5.5$ at mean maximum).

TABLE II
Mira variables in globular clusters

Cluster	Variable	Period (days)	M_v	Spectrum	Membership
47 Tuc	V1	212	−3.2	Me	+
	V2	203	−3.3	Me	+
	V3	192	−3.2	Me	+
NGC 6712	CH Sct	191	−3.6	Me	+
NGC 6637	V4	196		Me	+
	V10	?		Me	+
NGC 6388	V1	?		Me	+
	V4	?		Me	
	probably one other Mira				
NGC 6356	V3	?		Me	+
	probably two other Miras				
NGC 5927	Osborne	?		Me	
NGC 6553	V4	?		Me	

+ = radial velocity member

Table II summarizes the present position regarding Mira members of clusters. The three best known stars in this Table are the Miras in 47 Tuc, all with periods near 200 days. Following the work on the 47 Tuc variables (Feast and Thackeray, 1960), one star in NGC 6712 and two NGC 6637 were shown to have typical Mira spectra and to be radial velocity members of these clusters (Feast, 1967; Catchpole *et al.*, 1970). Of these three stars only two have had periods determined as yet. Both these periods are close to those of the 47 Tuc variables. Osborne (1968) located a slow variable near the centre of NGC 5927 and spectra obtained this season show it to be a typical Me Mira variable (unless otherwise stated the spectroscopic work mentioned in this paper was carried out at 140 Å mm^{-1} with the grating spectrograph and Carnegie Image Tube at the Cassegrain focus of the 74-in, (1.88 m) Radcliffe reflector). The period of this star is not yet known.

The discovery of Mira variables in clusters has recently been much accelerated by Lloyd Evans' work on V and I photography. He will be describing this work in collaboration with Menzies at this meeting, but I have taken the liberty of including some of his discoveries in Table II. The table lists several possible Miras he has selected in NGC 6356, 6388 and 6553 (in the last case the variable is one originally found by Thackeray (1955)). Table II also shows that where I have checked up on these stars spectroscopically they are found to have typical Me spectra. We obviously badly need periods for all these stars. The five known periods are close to 200 days. Dr Lloyd Evans' work suggests that the period of NGC 6553 V4 may be significantly longer than this – possibly 270 days.

One reason why a knowledge of the periods is of great interest is that the kinematics of Mira variables in the general field indicates a fairly good correlation of period and kinematic properties. Thus, in view of the small range of the known periods (191–212 days) we may inquire whether the clusters themselves form a rather tight group of similar objects.

TABLE III

Data on clusters containing Miras

Cluster	Morgan class	Deutsch/Kinman class	Kinman/Morgan CH/Hγ	H	Fe	ΔV	HB
47 Tuc		A	G3	G3	G3	2.15	Red
NGC 6712	V		G5			2.2	Red
NGC 6637	VII		G5	G8:	G2		
NGC 6388			G3	G3	G3		
NGC 6356	VI		G5	G2	G5	2.15	Red
NGC 5927			G2				
NGC 6553	VIII						

ΔV measures the height of the giant branch.
HB indicates which side of the RR Lyrae gap the horizontal branch is strongly populated.

Data on the clusters is summarized in Table III. Taking together the evidence of the colour magnitude diagrams (height of giant branch measured by ΔV, population of the horizontal branch (blue or red side)), Deutsch-Kinman classes, Morgan classes etc. it is clear that we are dealing with a group of metal rich globular clusters. These results are of course consistent with the kinematics of 200 day Miras in the general field, which indicate a very old population but not one as extreme as that of the very metal poor halo objects (Feast, 1963).

When 47 Tuc was first investigated spectroscopically one of the things that seemed remarkable was that the reddest non-variable stars were of spectral type M (up to about M2 at the latest) (Feast and Thackeray, 1960). It had not then seemed likely that globular cluster giants would be sufficiently metal rich and/or cool to show M type spectra. Over the years the presence of M type non-variable stars has generally been regarded as a unique property of 47 Tuc, though Stephenson (1961) found from objective prism work that there were some M type stars (M0 to M3) in the region of M17 which could be members of that metal rich cluster. M17 is not known to contain Mira variables and therefore does not appear in Table II. However there are a number of variables in the region including the long period irregular variable Z Sge which has a type of M4 according to Stephenson and $M_v \sim -1$ (Arp and Hartwick, 1971).

Are the other clusters in Table III sufficiently similar to 47 Tuc to show TiO bands in their non-variable red giants? A number of spectra have been taken to test this hypothesis. Table IV gives data on two red giants in NGC 6356. The photometry is

TABLE IV

(a) Non variable red giants in metal rich clusters

Cluster	Star	V	$(B-V)$	$(B-V)_0$	Spectral type
NGC 6356	5	15.35	2.05	1.55	M0–1
	46	15.35	2.22	1.72	M2
NGC 6637 (M 69)	IV-27 (inner)				M0 or 1
	III-42 (outer)				probably not M type
NGC 6553	TLE 24 (? variable)				M4

(b) Small amplitude red giants, in some metal rich clusters

NGC 5927	LF 4				M3
NGC 6637 (M 69)	V1				M2
	V3				probably not M type

from Sandage and Wallerstein (1960). Radial velocities of these stars are not known, but both lie on a well defined giant branch and are thus rather unlikely to be field stars. Star 46 is the reddest star in NGC 6356 measured by Sandage and Wallerstein and has the same spectral type as the latest red giants in 47 Tuc. Within the uncertainty introduced by the reddening correction to the NGC 6356 stars ($\pm 0^m.1$) the intrinsic colours of these M stars lie in the range of the 47 Tuc M giants (1.5–1.6). On this evidence NGC 6356 is quite similar to 47 Tuc.

In NGC 6637 (M69) Lloyd Evans and Menzies (1971) have shown that the photometry of Hartwick and Sandage (1968) needs substantial revision. However this photometry does at least allow us to pick out the stars at the red giant tip. The reddest giant not known to be variable (Hartwick and Sandage IV-27, inner zone) is of type M0 or M1. The next reddest giant (III-42 outer zone) does not show TiO bands with certainty (cf. Table IV). Evidently the non-variable giant branch of this cluster also penetrates into the early M types.

NGC 6553 is of special interest since its Mira may be of longer period than the others. It is certainly a metal rich cluster but otherwise rather little is known about it. However, Lloyd Evans finds a considerable number of stars with large $(V-I)$ colours in the region of the cluster. One of these stars (star 24 in Lloyd Evans' system) which may perhaps be a small range variable is of spectral type M4. This is definitely later than the latest non-variable red giants in 47 Tuc and marginally later than even the semiregular red variables in 47 Tuc. This is perhaps a tantalizing hint that longer periods are associated with greater metal richness and/or cooler giant tips.

It has long been recognized that the presence of TiO in the red giant variables in 47 Tuc produces a spurious blueing of $(B-V)$. Evidently one must anticipate the possibility of such effects in all the Mira-containing clusters. For these clusters, therefore, there is good reason to abandon $(B-V)$ and use colours in the infrared. Lloyd Evans and Menzies will be talking about their photographic work. Eggen (1972) has obtained (R, I) photometry which shows the important part this can play in globular cluster work. Wing will later be discussing results obtained with his narrow band system.

3. Small Amplitude Red Variables in Metal Rich Globular Clusters

Besides the three 200^d Mira variables, 47 Tuc contains five other red variables. The data on these stars are shown in Table V. Arp et al. (1963) noted that these eight variables seemed to divide into three groups with periods near 50^d, 150^d and 200^d and with amplitude and absolute magnitude at maximum increasing with period. It is not clear whether these groups are just a chance effect due to the small numbers of variables involved, or whether they are of significance for pulsation theory. In any case the spectral type becomes later with increasing period so that there are reasonable grounds for believing that stars evolve off the giant tip, moving to lower temperatures, with increasing amplitudes and periods. This conclusion is nicely illustrated by recent

TABLE V

47 Tuc red variables

Variable	Period	ΔV	V_{max}	M_v (max)	Max. type
5	45^d	0.4	11.5	-2.0	M2
6	47	0.6	11.3	-2.2	
7	58	0.5	11.4	-2.1	M2
4	165	1.8	10.9	-2.6	
8	155	1.7	10.9	-2.6	M2–3 (e)
1	212	4.4	10.2	-3.2	M2–3 (e)
2	203	3.7+	10.2	-3.3	M2–3 (e)
3	192	4.3	10.3	-3.2	M2–3 (e)

ω Cen TiO variables

RGO 320	?	0.5	12.4	-1.7	M2
V17	60	0.8	12.7	-1.4	M3–4
V6	100–120	1.2	12.2	-1.8	M4–5 (e)
V42	149	>3.3	11.2	-2.9	M2 (e)

	E_{B-V}	$V_0 - M_V$
47 Tuc	+0.03	13.4
ω Cen	+0.11	13.7

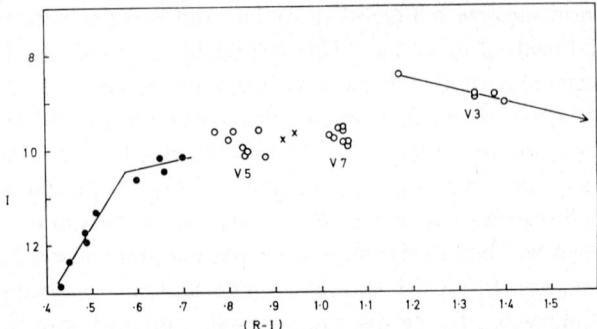

Fig. 1. $I,(R-I)$ diagram for 47 Tuc using data from Eggen (1972). Filled circles non-variables; crosses two small amplitude variables. Groups of open circles indicate the 45 day variable V5, the 58 day variable V7 and the 192 day Mira V3. In the case of V3 only points in the brighter part of the cycle are shown, other points lie outside the figure in the direction of the arrow.

(R, I) photometry by Eggen (1972) which is plotted in Figure 1. Included are two additional small amplitude variables noted by Eggen (one of them originally found by Arp et al. (1963)). Note how the variables lie to the red of the red giant tip with the Mira reddest of all. Comparison of this plot with a $V, (B-V)$ plot (Arp et al., 1963, Figure 10) shows the dramatic effect of TiO absorption on $(B-V)$. In this latter plot the three 50 day variables lie near the red giant tip but the 150^d and 200^d variables, which are cooler, are displaced to the blue. Even R and I may be somewhat affected by TiO and for some purposes there may be good reason to go further into the infrared. Glass has recently carried out some Johnson J, H, K, L photometry $(1.2-3.4\,\mu)$ in Pretoria which also clearly shows that the variables in 47 Tuc are cooler than the stars at the red giant tip.

All the other clusters with Mira variables also contain red variables with smaller amplitudes which could well be similar to the small amplitude variables in 47 Tuc. These stars promise to give useful information on evolution off the red giant tip, but so far the data is too fragmentary for a proper discussion. After 47 Tuc the most extensive discussion of non-Mira red variables in metal rich clusters is that on NGC 6712 by Rosino (1966) and by Sandage et al. (1966). Besides the Mira variable in this cluster, periods and light curves are known for 4 or 5 other red variables. The results resemble the 47 Tuc work at least to the extent that there are some SR variables of moderate amplitudes and periods of ~100 days. However, a meaningful discussion will hardly be possible until spectra or multicolour photometry have been obtained. Spectra are necessary to establish radial velocity membership, since the cluster is in a relatively rich field, and to help classify the variables. For example V2 has in the past been considered by one writer as an RV Tauri star and by another as an SRd variable, whilst it could well be an M type SR variable.

In the case of NGC 5927, Fourcade, Laborde and Albarracin (1966) have found several variables without determining periods or amplitudes. Lloyd Evans' work indicates that some of these have large $V-I$ colours and small amplitudes and may well be similar to the 47 Tuc SR's. In fact the spectrum of one of them (LF4) is of type $\sim$ M3, marginally later than the 47 Tuc SR's.

Another cluster containing Miras, NGC 6637 (M69) also has several small amplitude variables. One of these (V1) has a spectral type of $\sim$ M2. Again similar to the 47 Tuc SR's. The case of the small amplitude variables V3 in this cluster is interesting. It lies to the blue of the red giant tip in the $V, (B-V)$ diagram, and Lloyd Evans and Menzies (1971) called attention to its position above the giant branch in a pseudo $I, (V-I)$ diagram. TiO bands cannot be seen with certainty in recent spectra of this star. If it is a cluster member it is likely to be similar to several variables in other clusters which have been shown by Eggen (1972) to lie in the region of the giant tip but somewhat to the blue of the tip. Possibly these stars lie on loops in their evolutionary tracks extending from the giant tip to higher temperatures.

The discussion so far may be summarized as follows:
(1) There are a number of clusters containing Mira variables.
(2) The periods where known are near 200 days.
(3) These clusters are all metal rich.
(4) Where tests have been made it is found that the non-variable red giant branch penetrates into the M type region in these clusters.
(5) These clusters generally also contain red variables of shorter period, and smaller amplitude than the Miras. These variables are also of type M, but somewhat earlier than the Miras. Stars apparently evolve to the red of the giant tip as variables of increasing amplitude and period.
(6) In all these clusters TiO blanketing is likely to confuse the interpretation of the $V, (B-V)$ diagram.

4. Red Variables in ω Cen

Turning now to metal poor clusters, let us discuss work on ω Cen in some detail. There are about a dozen semiregular and irregular giant variables in ω Cen. Most of these were included in Martin's (1938) extensive early discussion. In the 1971 season Lloyd Evans extended his V, I work to ω Cen, and Dickens and I obtained spectra of variables that seemed interesting in this survey. I want to summarize this work (Dickens et al., 1972) as updated by further spectroscopic work this year. Briefly the main feature of the V, I work is that the red variables divide into two groups, one with $(V-I)$ values corresponding roughly to the tip of the giant branch, and one with much redder $(V-I)$ values. These redder stars are found spectroscopically to show TiO bands. Radial velocities have been obtained for many of the stars discussed. This is essential, at least in the case of the TiO variables, since we find both members and non-members in this group. M type non-members are not too much of a surprise since ω Cen is in a relatively rich field and covers a large area*.

Figure 2 shows the $I, (V-I)$ diagram for ω Cen. The giant branch is shown as dots from Brooke's (1969) work. Crosses are mean positions of non-TiO variables.

* RGO 4789, found variable by Dickens et al. (1972), has recently been found to be a radial velocity non-member.

These are quite small amplitude objects ($\Delta V \sim 0.5$ or less). The periods are in the range 70 to 124 days. It will be seen that these variables cluster around the giant tip. V164, a radial velocity member, falls below the red giant tip. It may or may not be significant that its period is ~ 37 days, shorter than any of the others. V167 falls well down the giant branch. However it was not found variable in our work and we do

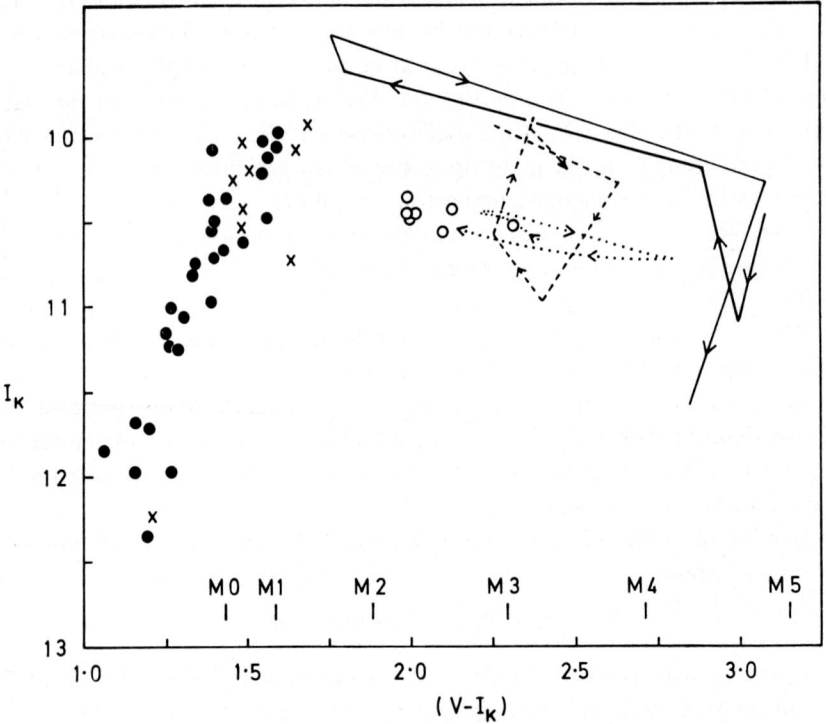

Fig. 2. $I, (V-I)$ diagram for ω Cen from Dickens *et al.* (1972) with the TiO variable RGO 320 (open circles) added. Filled circles are non-variable stars. Crosses are non-TiO variable (the faintest of these is V167 and the next faintest V164). The other TiO variables are V42 (full line), V6 (dashed line) and V17 (dash-dotted line). The non-variables are mostly from Brooke (1969).

not yet know if it is a member or not. These semiregular variables lying near the red giant tip and with spectra similar to red giants at the tip, presumably indicate the onset of instability at this point. They are presumably equivalent to the M type semi-regular variables of about the same period and amplitude in 47 Tuc, though occurring at a somewhat higher temperature (judging from Eggen's R, I measures and Glass' J, H, K, L results). Variables of this kind appear to be present in a number of metal-poor globular clusters.

The four variables whose paths in the $I, (V-I)$ diagram of Figure 2 are shown in detail are all radial velocity members of the cluster. There can be little doubt about this since the cluster has a high radial velocity*. Some data on these four variables are given in Table V and below. (a) V42 is affected by a close companion near minima. The star is generally classified as an SRd variable. It not known whether it moves out of the M types at maximum, but it shows TiO quite close to maximum. (b) V6 might be called a Mira except for its small amplitude. (c) RGO 320 was discovered to be variable by Lloyd Evans. (d) V17 was observed spectroscopically at monthly intervals in June, July, August 1971 and April, May, June, July and August 1972. The type from the TiO bands was always near M3 or M4. This coupled with the small amplitude and large $(V-I)$ values suggests that V17 and probably also V6 and RGO 320 always remain in the M types. It is obviously difficult to fix a precise mean point for these stars in the HR diagram. But even if we were to omit V42 from consideration because of its large amplitude, we would still find that in order to include the other TiO variables the giant branch would have to extend well to the red of the giant tip in $(V-I)$, and almost certainly also to bend down in I.

The most obvious, and most likely, interpretation of these results is that the evolutionary tracks extend to considerably cooler temperatures than we have been accustomed to think of for metal poor clusters. Can we rule out the possibility that these stars have temperatures similar to the non-TiO red variables, and that the strong TiO bands indicate abundance anomalies of an unsuspected kind? The large $(V-I)$ colours are probably no clear cut evidence against this since these colours are strongly affected by TiO blanketing. There are, however, at least two reasons for rejecting this hypothesis.

(1) The $V, (B-V)$ diagram was discussed by Dickens *et al.* (1972). Here again the non-TiO variables cluster around the red giant tip. The TiO variables have similar $(B-V)$ colours but lie below them. For the M stars the effects of TiO blanketing are marked. Making approximate blanketing corrections to V and $(B-V)$ (cf. Smak, 1966) it is found that the TiO variables move up to about the same V as the non-TiO variables and out to $(B-V) \sim 2.0$. This certainly suggests that the presence of TiO is a real indicator of low temperature.

(2) Glass (to be published) has obtained J, H, K, L photometry of a number of stars at the red giant tip in ω Cen, as well as some non-TiO variables and the TiO variables V17 and V6. A $J, (J-K)$ plot shows that the TiO variables have much larger $(J-K)$ values, indicating them to be much cooler than the other stars. After correction for a small amount of interstellar reddening, the $(J-K)$ colours of V17 and V6 agree with their TiO types (using the calibration of $(J-K)$ with spectral type given by Lee (1970)). Thus either these stars have compositions similar to ordinary field M giants, which is a little surprising for stars in a metal-poor globular cluster, or else the $(J-K)$, TiO-type relation is insensitive to metal abundance.

It is interesting to compare the TiO variables in ω Cen with those in 47 Tuc. The most conspicuous difference is that whilst the TiO variables in ω Cen lie below the giant

---

* Figure 2 contains an extra TiO variable compared with Figure 2 of Dickens *et al.* (1972). This is RGO 320 noted in that paper as a small amplitude variable and now found to be a radial velocity member (radial velocity $+226$ km s^{-1} from a 50Å mm^{-1} Carnegie Image Tube Spectrum).

branch in the $V, (B-V)$ diagram, the 47 Tuc variables lie on or slightly above the giant branch at mean light. This is presumably due either to different shapes to the giant branches in the two clusters and/or to TiO blanketing being different in the two clusters at the same temperature.

Table V compares the variables in 47 Tuc and ω Cen. The 160 day variables in 47 Tuc have $M_v = -2.6$ at maximum, whilst V42 ω Cen has M_v (maximum) $= -2.9$, but since this star has a much bigger amplitude it is fainter at mean light than the 47 Tuc stars. The 50 day variables in 47 Tuc have $M_v = -2.1$ at maximum. V17 with a similar period and amplitude has $M_v = -1.4$ at maximum and a later spectral type. RGO 320 ($M_v = -1.7$ at maximum) is also fainter than the 47 Tuc variables. These differences could be reduced or possibly eliminated by increasing the modulus of ω Cen by $\sim 0\overset{m}{.}5$. This is perhaps not out of the question. If there are real differences in absolute magnitude and other properties between the TiO variables in 47 Tuc and ω Cen, then one is faced with the unpleasant prospect that both 47 Tuc-type and ω Cen-type M variables may be present in the general field, and that these cannot be readily distinguished from one another on the basis of photometry or low dispersion spectroscopy.

5. The SRd Variables

In the general field there are about 15 to 20 rather similar stars which are generally grouped together under the name SRd variables. They have peculiar spectra, probably indicative of metal deficiency. They are generally classified Fp, Gp or Kp and usually they show TiO bands towards minimum. One of their chief characteristics is the presence of strong hydrogen emission lines near maximum. Generally their periods are in the range 75 to 120 days and their amplitudes between 1 and 4 mag. Their radial velocities show them to belong to a halo-type population, and it has been proposed that they form an extension of the main Mira variable sequence to periods shorter than 200 days into the halo population (Feast, 1965; Preston, 1967). Obviously as such we may well expect to find them in globular clusters. Table VI shows several variables in globular clusters which are probably SRd variables on the basis of Joy's (1949) spectroscopic work. They show hydrogen emission and in at least one case TiO bands near minimum. It may be of significance that their amplitudes are low compared with field SRd's, though there is one in the field with an amplitude less than 1^m (UW

TABLE VI
SRd variables (Joy)

NGC	M	Variable	Period	Amplitude	Spectrum	Morgan class	ΔV
6656	22	5	?	0.8	G0–G6 (e)		
6656	22	8	67 ±	0.7	G2–G5 (e)	II	2.5
5272	3	95	103	0.7	G4–M4 (e)	II	2.64

ΔV = Height of giant branch.

Lib $\Delta m \sim 0.7$). Some of the other variables observed by Joy could be SRd's though they did not have hydrogen emission when he observed them. However these latter stars may also be similar to the non-TiO variables in ω Cen, which do not classify as SRd's as their amplitudes are too small and they do not have hydrogen emission.

As mentioned in the last section, V42 ω Cen seems best classed with the SRd's. Its period is, however, long compared with typical SRd's in the field or others known in globular clusters. V42 ω Cen seems to bridge the gap between 100 day SRd's and 200 day Miras. We know that the kinematics of the field stars in the main Mira-SRd sequence are correlated with period and this presumably indicates some correlation of metal abundance with period. We have, however, little idea of just how close this correlation is, though we have seen that Miras with periods close to 200 days occur in a group of rather similar metal-rich clusters. V42 ω Cen provides an interesting problem in this connection. Table VII compares this variable with the bright Mira variable

TABLE VII

Comparison of V42 ω Cen and S car

Star	Period	Amplitude	Spectrum	Radial velocity (km s^{-1})
V 42	149.4	> 3.3	~ M2e	+261
S Car	149.5	mean 2.9 mean 5.4	K7e–M4e	+289

S Car. The periods and amplitudes are very similar. Like ω Cen, S Car is a high velocity object, which presumably indicates membership of the halo population. Hain (1969) has made a detailed study of S Car. Despite the very high velocity, the metal deficiency is only moderate ($\sim 1/10$ solar). Dickens (this conference) has summarized the present position regarding abundances in ω Cen. Whilst the ultraviolet excesses of the giants indicate metal abundances of $\sim 1/50$ solar, a preliminary spectroscopic analysis of Fehrenbach's bright member suggests an abundance of $\sim 1/15$ solar. Is the value derived for Fehrenbach's star the true value for the cluster as a whole? If not, could it at least apply to V42? It should not be impossible to answer these questions and this would greatly advance our knowledge of red variables in globular clusters.

Red variables are likely to become of increasing importance in attempts to understand the Magellanic Clouds. I want now in conclusion to discuss the brightest red variable in the SMC globular cluster NGC 121 in an attempt to see how it is related to the red variables we find in globular clusters in the Galaxy. The variables in NGC 121 were first investigated by Thackeray (1958) and later by Tifft (1963). The photometric properties of the brightest variable V1 are given in Table VIII. A recent spectrum taken near maximum light shows a fairly smooth continuum with no sign of TiO bands but with strong hydrogen lines in emission. The star thus almost certainly classes as an SRd variable. In Table VIII it is compared with two other SRd's, V42 ω

Thackeray, A. D.: 1955, see H. B. Sawyer-Hogg, *Publ. David Dunlap Observatory* **2**, No. 2.
Thackeray, A. D.: 1958, *Monthly Notices Roy. Astron. Soc.* **118**, 117.
Tifft, W. G.: 1963, *Monthly Notices Roy. Astron. Soc.* **125**, 199.

DISCUSSION

Buscombe: (1) Is it not important to have quantitative measurements of equivalent width for (e.g.) TiO bands, from which abundance and temperature effects can be disentangled from broad-band photoelectric colour indices?
(2) Observers should always quote the date of colour measures for any type of variable star.

Schwarzschild: Dr Feast has I believe given us a remarkably systematic review of the variables at the top of the red giant branch in globular clusters. The data he gave seem to me most encouraging for the view that a star climbing up the red giant branch for the second time first becomes pulsationally weakly unstable, then as it rises, more and more unstable (longer periods, larger amplitudes), until during a pulsation the radius becomes dangerously large (very red minima), which finally should lead to the ejection of a planetary nebula.

Dickens: In connection with Prof. Swarzschild's remarks concerning the possibility of mass loss occurring in these very cool variables, the absorption lines in the spectra of some of the red variables in ω Cen, in particular V42, show a 'washed out' appearance, presumably similar to what has been observed in some field red variables. One possible explanation for this effect would be some absorption from particles high in the atmosphere or perhaps in a shell which has been ejected from the star.

Jones: Could you comment on the fact that Kinman did not record TiO in his spectra of giants in ω Cen?

Feast: Kinman had, I think, only one spectrum of a TiO variable (V42) near maximum. TiO is not strong on the plate, but if I remember correctly, it is rather over-exposed in the relevant spectral region. Generally I think Kinman kept away from the extreme giant tip in his work.

Jones: My radial velocities of M stars in the SMC reveal only one candidate for an analog to the ω Cen M stars, which is geographically in the halo. L² Pup is a much more obvious analog.

Wing: The absolute magnitudes of the Mira variables in 47 Tuc can hardly be fainter than $M_{bol} = -5.0$, to judge from the M_V's and spectral types you tabulated. With regard to their apparent V magnitudes, they may be brighter than giant-branch stars of the same $B-V$ color some of the time, as you commented, but most of the time they are below the giant branch. At minimum they reach $V = 15$ or 16 without much change in $B-V$.

Feast: The V magnitudes plotted in the diagram were time-averaged intensity values.

van den Bergh: You suggest that we might find a galactic counterpart to V1 in the SMC cluster NGC 121 *if* we look hard enough. But is it not true that such very red objects would have been found if they had been located in galactic globular clusters?

Feast: It does seem unlikely that there are any equivalent stars in galactic globular clusters, though only a few red giant variables in galactic globular clusters have $(B-V)$ colour curves.

A PRELIMINARY INVESTIGATION OF PERIOD CHANGES FOR W VIRGINIS STARS IN GLOBULAR CLUSTERS

CHRISTINE M. COUTTS

David Dunlap Observatory, University of Toronto, Richmond Hill, Ontario, Canada

Investigations in recent years have shown that there may be two mechanisms which place stars in the W Virginis instability region (Kraft, 1972). The variables with periods less than 8 days seem to be in the stage of 'above horizontal branch' evolution discussed by Strom *et al.* (1970). The longer period group apparently results when thermal instabilities in the helium burning shell of an asymptotic red giant branch star cause it to loop to the left in the HR diagram. This longer period group has been investigated by Schwarzschild and Härm (1970) and Mengel (1972). The present study has been undertaken to see if there are any notable differences between the period changes of variables belonging to the two groups.

Period changes for the variables in M13 have already been discussed by Osborn (1969). The periods of the W Virginis stars in M5 have been studied by Coutts and

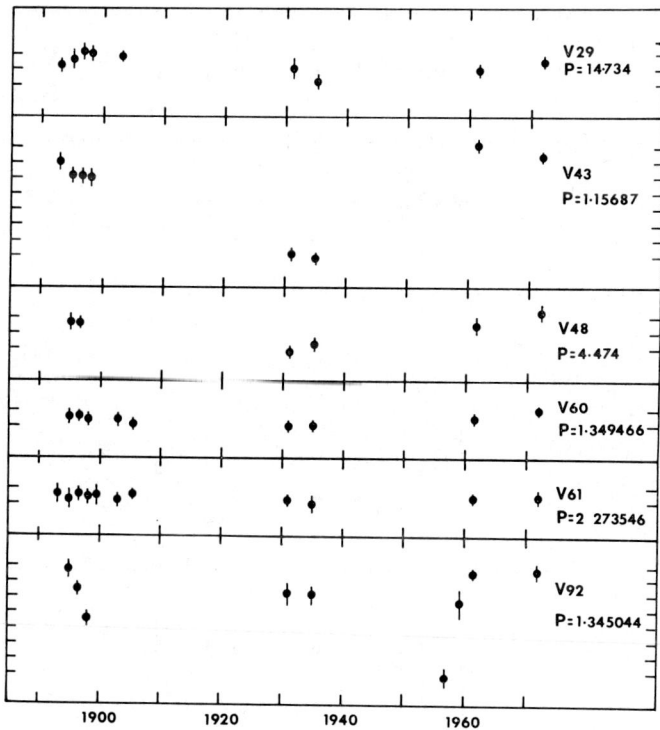

Fig. 1. Phase-shift diagrams for the variables in ω Centauri. The spacing of the marks on the vertical axis is 0.2 of the period. The date in years is indicated along the horizontal axis. Error bars indicate probable errors.

Sawyer Hogg (1972a, 1972b). In the present investigation, phase-shift diagrams have been plotted for the W Virginis stars in three other clusters: ω Centauri, M14 and M2.

The material for ω Centauri has been taken from the published work of Bailey (1902), Sawyer (1931), Martin (1938), Eggen (1961), Dickens and Carey (1967), and Wilkens (1967). In addition, the author obtained a series of photographs of ω Centauri with the David Dunlap 24-in. telescope at Cerro Las Campanas, Chile in May and July, 1972. These phase-shift diagrams are shown in Figure 1. The diagrams for M14 (Figure 2) are based on the material published by Sawyer Hogg and Wehlau (1966, 1968), and Demers and Wehlau (1971). Those for M2 (Figure 3) are from Sawyer (1935) and from Poole (1968), a Master's thesis based on Sawyer Hogg's collection of plates at the David Dunlap Observatory.

Table I summarizes the results of the investigations of period changes of W Vir-

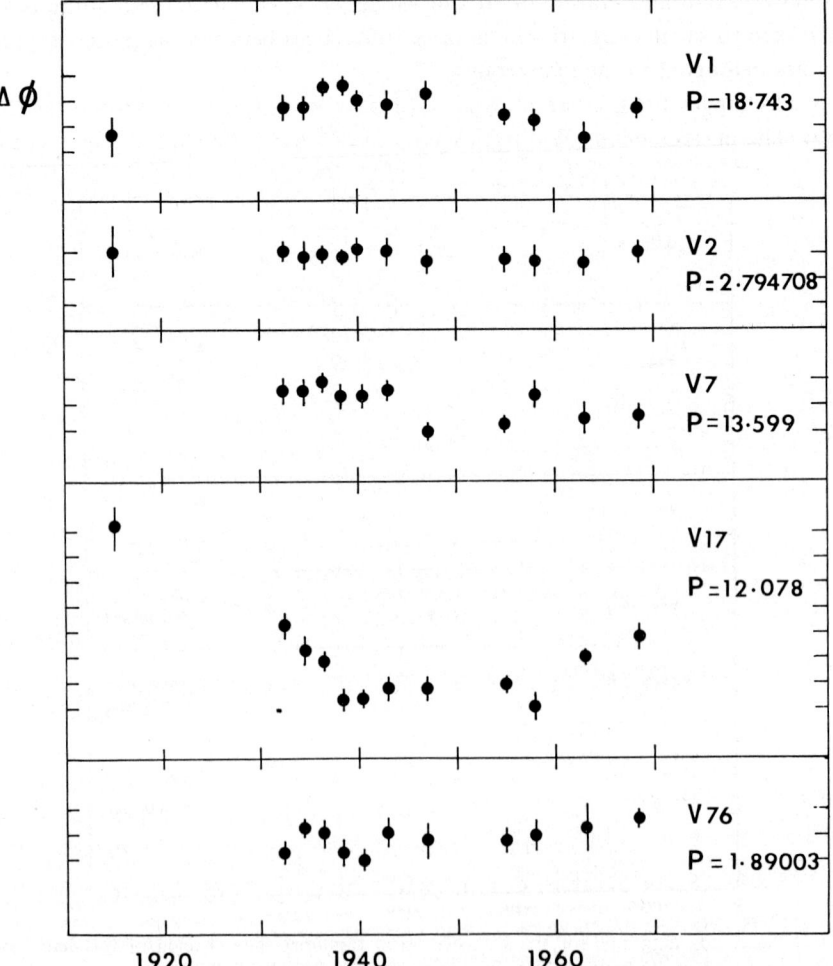

Fig. 2. Phase-shift diagrams for the variables in M14.

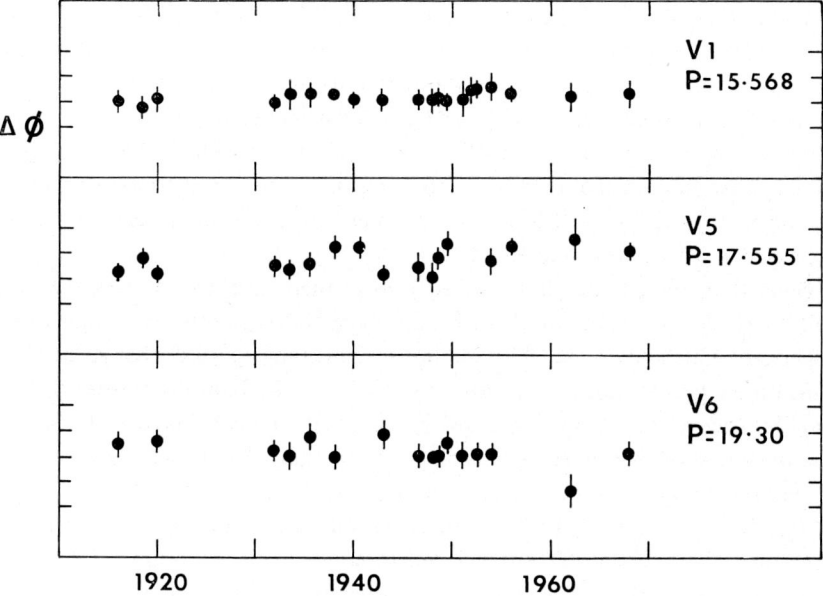

Fig. 3. Phase-shift diagrams for the variables in M2.

TABLE I

Summary of the period changes

Cluster	Variable	Period	Remarks about period change
ω Cen	29	14$^\text{d}$734	small fluctuations
	43	1.15687	fluctuations
	48	4.474	increasing
	60	1.349466	increasing
	61	2.273546	constant
	92	1.345044	fluctuations
M5	42	25.738	constant
	84	26.42	fluctuations
M13	1	1.458997	probably constant
	2	5.110939	increasing
	6	2.112867	probably constant
M14	1	18.743	fluctuations
	2	2.794708	constant
	7	13.599	fluctuations
	17	12.078	increasing
	76	1.89003	probably constant
M2	1	15.568	constant
	5	17.555	constant
	6	19.30	small fluctuations

ginis stars in the five globular clusters, ω Cen, M5, M13, M14 and M2. Of the 19 stars studied, 10 have periods less than 8 days and 9 have periods longer than 8 days. In the shorter period group, 5 have constant periods, 3 have increasing periods and 2 have periods which have fluctuated during the last 60 years (approximately). (If the phase-shift diagram can be fit by a straight line, we have assumed that the period has remained constant over the interval of observations. This is not necessarily the case). In the longer period group, 3 have constant periods, 1 has an increasing period and 5 have periods which have fluctuated.

It seems that among the short period group, there is a greater tendency for the periods to remain constant or to increase, while the variables with longer periods have periods which fluctuate. The minimum detectable period change in 60 years depends on the period itself and is given by $\Delta P = 10^{-5} P$. Thus the shorter the period, the smaller the period change that can be detected. It may therefore be significant that half of the short period stars studied have constant periods, while over half of the longer period variables have fluctuating periods. At this point, it is difficult to conclude anything about the tendency for the shorter period stars to increase their periods. This result comes mainly from ω Centauri and it can be seen in Figure 1 that there are large intervals when the W Virginis stars in this cluster were not studied.

Period changes for some W Virginis stars in the field have been studied by Kwee (1967). The longer period stars tend to have fluctuating periods, similar to those in the clusters. However, of the 3 short period stars, one has a decreasing period, one a fluctuating period and one a constant period, over an interval of about 40 yr.

At the present time, it appears that the W Virginis stars with periods greater than 8 days have periods which either remain constant or fluctuate randomly. The variables with periods less than 8 days have not yet been studied adequately. When more observations are available for ω Centauri, we hope to understand these stars more fully.

References

Bailey, S. I.: 1902, *Harv. Ann.* **38**, 1.
Coutts, C. M. and Sawyer Hogg, H.: 1972a, *Bull. Am. Astron. Soc.* **4**, 217.
Coutts, C. M. and Sawyer Hogg, H.: 1972b, in press.
Demers, S. and Wehlau, A.: 1971, *Astron. J.* **76**, 916.
Dickens, R. J. and Carey, J. V.: 1967, *Roy. Observ. Bull.*, No. 129.
Eggen, O. J.: 1961, *Roy. Observ. Observ. Bull.*, No. 29.
Kraft, R. P.: 1972, in A. G. Davis Philip (ed.), *The Evolution of Population II Stars*, Dudley Obs. Reports No. 4.
Kwee, K. K.: 1967, *Bull. Astron. Inst. Neth. Suppl.* **2**, 97.
Martin, W. Chr.: 1938, *Ann. Sterrew. Leiden* **17**, pt. 2, 1.
Mengel, J. G.: 1972, IAU Colloq. No. 21, Toronto.
Osborn, W.: 1969, *Astron. J.* **74**, 108.
Poole, L.: 1968, Master's Thesis, University of Toronto.
Sawyer, H. B.: 1931, *Harv. Circ.*, No. 366.
Sawyer, H. B.: 1935, *Publ. Dom. Astrophys. Obs.*, Victoria **6**, 265.
Sawyer Hogg, H. and Wehlau, A.: 1966, *Publ. David Dunlap Obs.* **2**, 451.
Sawyer Hogg, H. and Wehlau, A.: 1968, *Publ. David Dunlp Obs.* **2**, 493.
Schwarzschild, M. and Härm, R.: 1970, *Astrophys. J.* **160**, 341.
Strom, S. E., Strom, K. M., Rood, R. T., and Iben, I.: 1970, *Astron. Astrophys.* **8**, 243.
Wilkens, H.: 1967, *Mitt. veränderl. Sterne* **4**, 93.

DISCUSSION

Wesselink: Have you noticed any changes in form of light curves?

Coutts: There seems to be, but the results are difficult to obtain because of differences in scales.

Wehlau: A drop in amplitude over the years is seen in the Toronto observations for many variables in M14 and may be due to the gradual increase in sky background at Toronto.

Belserene: I have looked at V 60 and V 61 in ω Cen and have the impression that there are real variations in the shape of the light curves. Also, a very rapid period increase for V 92 fits many of the epochs. I haven't looked at them all. The rate is about 10 cycles per million years.

MEAN MAGNITUDES AND COLOURS OF SIX CEPHEIDS IN THREE RED GLOBULAR CLUSTERS OF THE LARGE MAGELLANIC CLOUD

SERGE DEMERS*
Laurentian University, Sudbury, Ont., Canada

Abstract**. Photographic B and V light curves are determined for six variables, with periods longer than one day, in and near NGC 1751, NGC 1953, and NGC 2121. New photoelectric sequences are used to calibrate the plates. The mean magnitudes and colours of these variables are similar to the magnitudes and colours of Classical Cepheids of the same period. The photometric properties of these variables are unlike Population II Cepheids in the Galaxy but are comparable to field Cepheids of the Large Magellanic Cloud.

DISCUSSION

Feast: Can one estimate the possibility of these Cepheids being Magellanic Cloud field stars (not cluster members)?

Demers: This has been done by Hodge and Wright. They compared the number of variables near their red clusters and in some fields of the LMC. They estimated that more variables were near the red clusters per unit area than in the outer fields. That is the major reason why they believe in the cluster membership of these variables.

---

* Visiting Astronomer, Cerro Tololo Inter-American Observatory.
** Complete details of this work will be published elsewhere.

RED VARIABLE STARS IN METAL RICH GLOBULAR CLUSTERS

T. LLOYD EVANS

Radcliffe Observatory, Pretoria; Royal Observatory, Edinburgh

and

J. W. MENZIES

Radcliffe Observatory, Pretoria; University Observatory, Oxford

1. Introduction

The globular clusters contain sufficiently large numbers of stars to permit a systematic study of the intrinsically rare variable stars which lie near the tip of the red giant branch. The position of the smaller amplitude stars in the colour magnitude diagram is of particular interest. Eggen (1972) has published photoelectric observations of such stars in several globular clusters, most of intermediate or low metal abundance. The mean colour of 14 stars in 5 clusters is $(V-I_K) = 1.40$, with a spread from 1.60 to 1.12 (or 0.75 if V8 in M22 is of this type) which Eggen regards as indicating a range of temperature. The red variable stars in the metal rich globular cluster 47 Tucanae are much redder and show a considerable range of colour.

It is of interest to know whether the colours of the red variable stars in globular clusters depend on the metal content of the cluster. The colour magnitude diagrams on the V, I_K system have been found for a sample of metal-rich clusters containing red variable stars, many of which are new discoveries.

I_K is the infrared magnitude on the system of Kron and Smith (1951). The clusters were selected for a small (negative) value of Q (van den Bergh, 1967), late spectral type, and weak or nonexistent blue horizontal branch, as well as observational convenience and the availability of supporting data. Details of the observations, made with the 1.88 m reflector at Pretoria, will be published elsewhere.

2. The Variable Stars

The available plates have been searched for variable stars. The number of plates, most of which are on the V system, is quite small and the search cannot be considered exhaustive. The high proportion of variables near the red giant tip shows that in certain clusters a substantial proportion of such variables have been detected, however. Some variables will have escaped detection because they were always at the same phase, always below the plate limit, or because they happened not to vary much over the period of the observations; another major problem in detection is that caused by the very crowded fields involved.

All suitable plate pairs were blinked at the Radcliffe Observatory or at the University of St. Andrews Observatory. Many of these plates, and others which were not taken specifically for the study of variable stars, were measured with iris photometers.

TABLE I
Globular clusters

NGC	Epochs	Seasons	$E(B-V)$	$E(V-I)$	Q
104	5	2	0.07	0.08	−0.26
5927	11	4	0.60	0.72	−0.13
6171	2	1	0.28	0.35	−0.35
6352	8	4	0.30	0.36	−0.09
6356	3	1	0.39	0.47	−0.24
6388	3	2	0.36	–	−0.20
6553	23	6	0.71	–	−0.11:
6637	11	3	0.20	–	−0.21
6712	2	2	0.48	0.58	−0.31
6723	Many	6	0.00	0.00	−0.28

TABLE II
Variable stars

Cluster	Star	ΔV	A_V	$(V-I_K)$	Remarks
104	V1	3+	4.4	–	Mira $P=212^d$
	V2	3+	3.7	–	Mira $P=203^d$
	V3	5:	4.3	–	Mira $P=192^d$
	V4	1.2	1.8	+3.16	$P=165^d$
	V5	0.3	0.4	+1.92	M2III: $P=45^d$
	V6	0.2	0.6	+1.87	$P=47^d$
	V7	0.5	0.5	+2.34	M2: $P=58^d$
	V8	0.7	0.7	+2.46	M2-3II: $P=150^d$
	V11	0.6	0.8 (m_{pg})	+2.69	HV813 W12
	V13	0.5	–	+1.83	Wilkens[a] W173
	W300	0.3	–	+2.09	Arp *et al.* (1963)
	R18	0.2	–	+1.94	M Brooke (1969)
	W81	0.1	–	+2.00	Eggen (1972)
	L168	0.3		+2.13	[b]
	R10	0.6		+1.60:	K311
	A1	0.8		+2.26	
	A2	1.0		+2.86:	
	A4	0.5		+2.05	
	A6	0.6		+1.82	
	A8	0.3		+1.78	
	A9	0.5		+2.13	
	A13	–		+2.48	Crowded
	A18	0.3		+2.48	
	LR5	0.4		+2.31	
5927	FL4	0.6		+3.93	Non-member?
	FL14	0.6		+3.13:	Crowded
	V1	3+		–	Mira $P\sim 300^d$
	V3	0.7		+3.40	
	V6	0.5		+3.44	
	V7	0.6		+3.05	
	V8	0.6		+3.37	
	V9	0.6		+3.64	
	V10	0.9		+3.69	
	L43	0.4		+3.11	

Table II (Continued)

Cluster	Star	ΔV	A_V	$(V-I_K)$	Remarks
	L17	0.4		$+4.51$	Non-member?
6171	SK217	0.4		$+2.22$	
6352	FL4	0.1		$+2.92$	Non-member?
	L36	0.5		$+3.51$	Non-member?
	HH113	0.6		$+3.04$	
6356	V1	1.5+		$\gtrsim +3.6$	Mira
	V3	2+		$\gtrsim +3.0$	Mira Feast (1972)
	V4			1.6 (mpg) $+4.2$	Mira Non-member?
	SW34	0.3:		$+2.50$	
	SW72	0.4		$+3.14$	
	SW30	0.4		$+2.68$	
	SW46	0.4:		$+2.90$	Probable
6388	V1			–	Mira Feast (1972)
	V2			$> +150$	Mira
	V3			$+34$	Bright
	V4			$> +100$	Mira
	V6			-5	
	V7			$+36$	
	V8			$+11$	
	V10			-79	Bright, not very red.
	V11			$+34$	
6553	V4			–	Mira $P=270^d$?
	T23			–	V5
	5			$+98$	T11
	A1			$+150$	
	A2			$+158$	
	3			$+152$	
	6			$+181$	
	7			$+129$	
	13			$+174$	
	14			$+173$	
	24			$+152$	
	33			$+156$	
6637	V1			$+26$	
	V3			-25	Bright
	V4			$(+178)$	Mira $P=196^d$
	V10			$(+128)$	Mira $P=195^d$
	II-37			$+80$	$1' < r < 2'$
	III-43			$+1$	$1' < r < 2'$
	IV-11			$+79$	$r < 1'$
6712	V2	0.9	1.4	$+2.38$	AP Sct, $P=105^d$
	V7	1.5	5.0	$\geq +2.2$	CH Sct, $P=190^d$ Me
	V8	–	1.9	$+2.91$	$P=117^d$
	V10	0.2	0.4	$+2.70$	
	V15	0.2	1.0:	$+3.00$	
	L18	0.3		$+2.40$	Probable
6723	V25	0.9		$+1.96$	$(V-I_K) = 1.86$ pe
	V26	0.9		$+1.62$	$(V-I_K) = 1.96$ pe

[a] See Fourcade and Laborde (1966).
[b] Marked VII on finding chart, in error.

Notes to Table II

W	Wildey (1961) in NGC 104.
R	Feast and Thackeray (1960) in NGC 104. Spectral types.
FL	Fourcade and Laborde (1966) in NGC 5927 and NGC 6352.
SK	Sandage and Katem (1964).
HH	Hartwick and Hesser (1972).
SW	Sandage and Wallerstein (1960).
T	Thackeray (private communication). The numbering is a provisional one from a search by Thackeray for variables in a 40′ × 30′ field centred on NGC 6553.

NGC 6637: see Hartwick and Sandage (1968), Rosino (1962), Catchpole *et al.* (1970), Lloyd Evans and Menzies (1971).

VI in NGC 5927 is Osborne's (1968) variable; otherwise stars identified by V, above a ruled line, are numbered according to Hogg (1955). Identifications below the line are assigned by the authors.

The values given for $(V-I_K)$ for NGC 6388, 6553 and 6637 are the mean difference of iris readings, $S_I - S_V$.

Stars denoted as non-members are considered to lie outside the cluster.

Criteria for deciding on the reality of variability included agreement of blink microscope and iris photometer results and confirmation from additional plates taken on the same nights. Judgement had to be exercised especially where the star concerned lay near the centre of the cluster. Magnitude differences of $0^m.2$ might be considered real in favourable cases but the detection threshold was larger in such compact clusters as NGC 6356 and 6388.

Table I lists for each cluster the number of epochs (dark of moon periods) for which observations are available; number of observing seasons; reddening adopted in plotting the CM diagram; and Q (van den Bergh, 1967). Table II contains the star identification, from published lists where available (those above a horizontal line being already identified in the literature); light range ΔV; amplitude A_V if already known from an adequate series of observations; $(V-I_K)$; remarks.

Finding charts are given for NGC 6388, 6553 and the central region of NGC 104.

TABLE III

The variable stars in metal rich clusters

NGC	Q	Sp	CM	RR	L+SR	M
6352	−0.09	–	R	–	1	0
6553	−0.11	–	R	1	11	1
5927	−0.13	G2	R	0?	10	1
6388	−0.20	G3	R	–	6	3
6637	−0.21	G5	R	0?	5	2
6356	−0.24	G5	R	0?	4	2(+1)
104	−0.26	G3	R	3	21	3
6723	−0.28	G2	$B \gtrsim R$	27	2	0
6712	−0.31	G4	$R > B$	8	5	1
6171	−0.35	G0-1	$R > B$	21	1	0
10				52	66	14

R: Red stub horizontal branch
B, R: relative strengths of blue and red horizontal branches
RR, L+SR, M: numbers of RR Lyrae, irregular and semi-regular, and Mira variables, respectively.

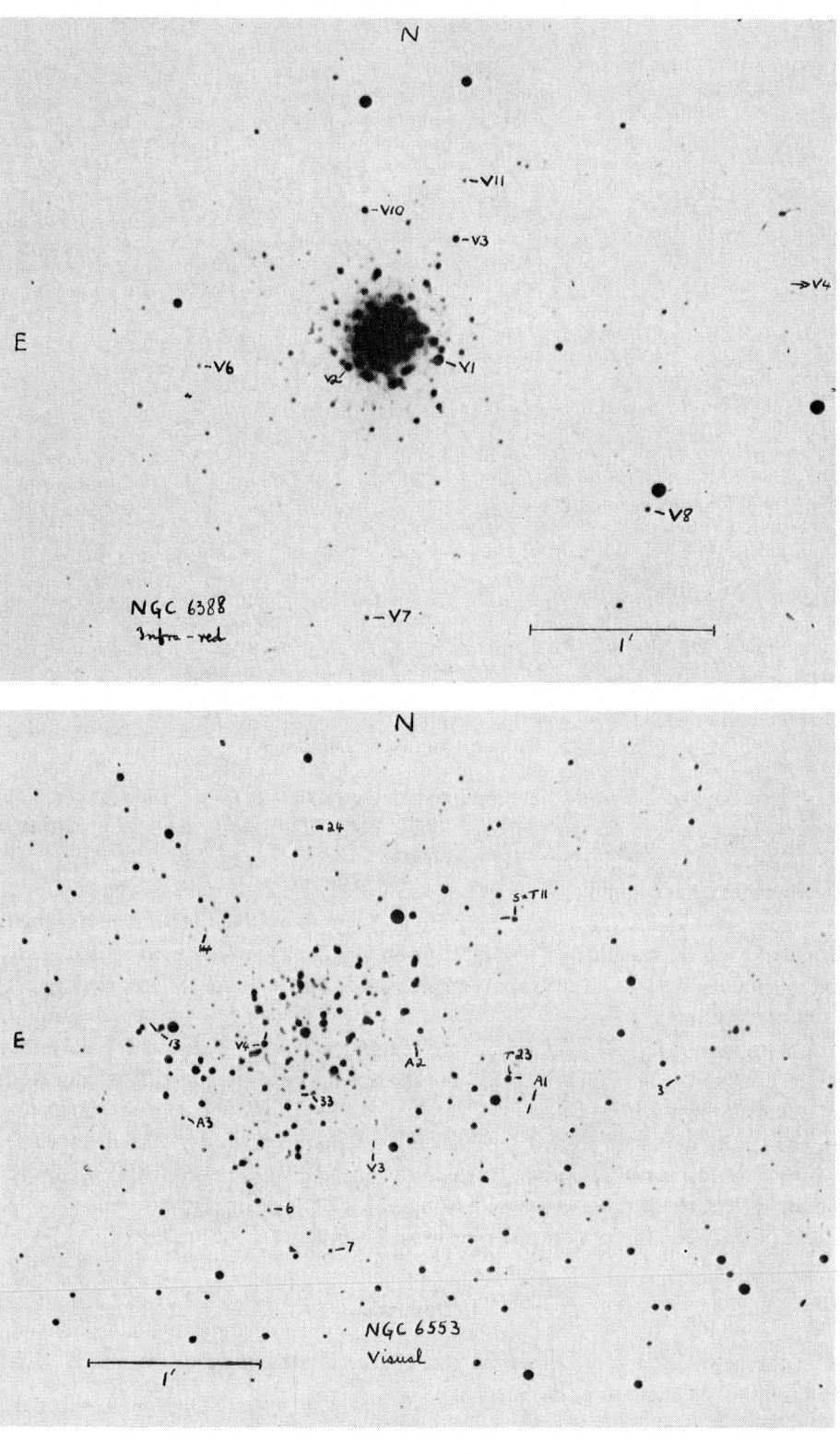

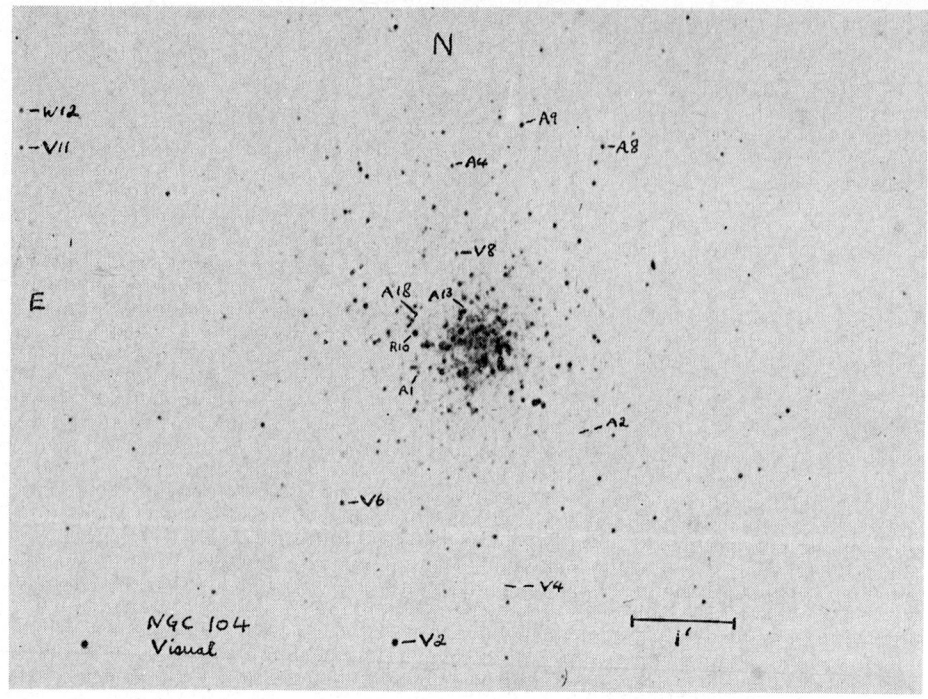

3. Colour Magnitude Diagrams

Colour magnitude diagrams were obtained on the I_K, $(V-I_K)$ system with calibrations as follows: V, I photoelectric sequence: NGC 104, 6723. Transformed UBV sequence: NGC 5927, 6171, 6352, 6356, 6712 and 6723.

No calibration available: NGC 6388, 6553, 6637.

Figure 1 contains the colour magnitude diagram of NGC 104, the stars with $2' < r < 8'$ and those with $r < 2'$ (where crowding and background light caused errors) being shown separately. The range of V magnitude measured on (usually) eight plates is plotted against colour.

The colour-magnitude diagrams for the remaining clusters are given in Figures 2–5. Those for NGC 6388, 6553 and 6637 are the uncalibrated pseudo-colour-magnitude diagrams, S_I against $S_I - S_V$.

The selection of stars plotted generally follows that of previous workers, especially with regard to the radius within which cluster members were expected to occur. The use of V, I plates of short exposure has in a few cases permitted measurements to be made of stars close to the centre or crowded on B plates.

4. Discussion

47 Tucanae (NGC 104) provides the best chance of studying the systematics of red variables. The following points emerge.

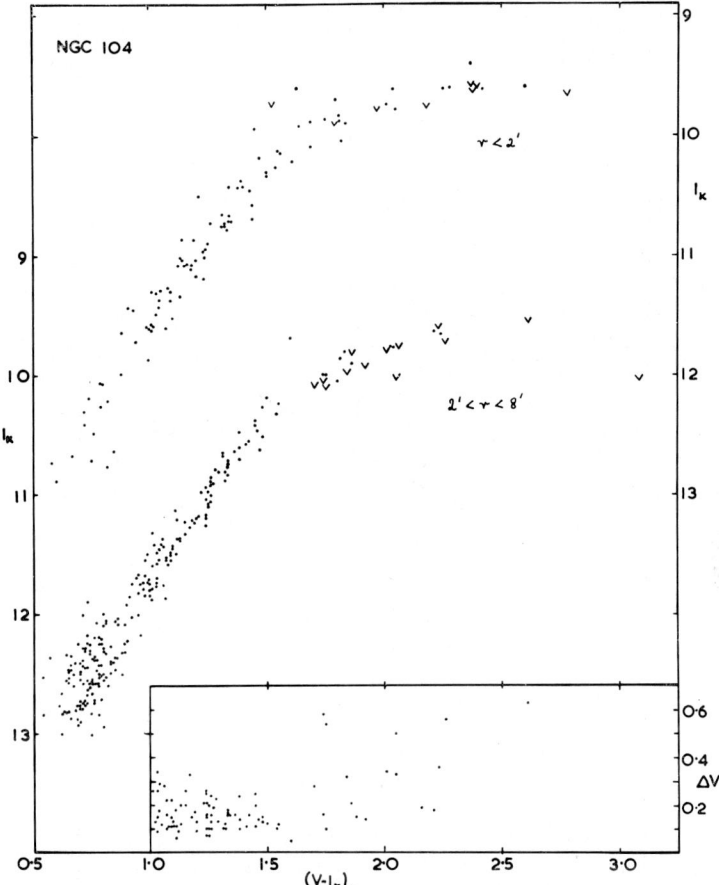

Fig. 1. *Top*: Stars within 2′ of the centre of NGC 104. All stars with $(V-I_K)_0 > 1.50$ are plotted but only uncrowded stars with smaller $(V-I_K)$. I is uncorrected for reddening. *Middle*: All uncrowded stars with $2' < r < 8'$. In both plots all stars considered variable by others are plotted as V, but only new cases considered definitely variable ($\Delta V \gtrsim 0\overset{m}{.}3$) are so indicated though other stars are suspected to vary. *Lower*: The range ΔV of V magnitudes found on 8 plates for stars with $(V-I_K)_0 \geqslant +1.00$ from the middle plot.

(1) Variability sets in at $(V-I_K) = 1.7$.

(2) The proportion of stars which are variable, and the amplitude of variation, both increase with colour. The plot of light range against colour suggests that all stars with $(V-I_K) > 1.8$ may be variables of small amplitude.

(3) There may be a gap, or interval of low star density, from $(V-I_K) = 1.5$ to 1.7.

(4) The colours of stars of known spectral type (Feast and Thackeray 1960) indicate that $(V-I_K) = 1.7$ corresponds to spectral type M0.

(5) The apparent downward trend in I_K of the red giant tip may be spurious, as the average magnitude and colours plotted represent the mean of isolated measurements of magnitude instead of the average intensity which would be preferable.

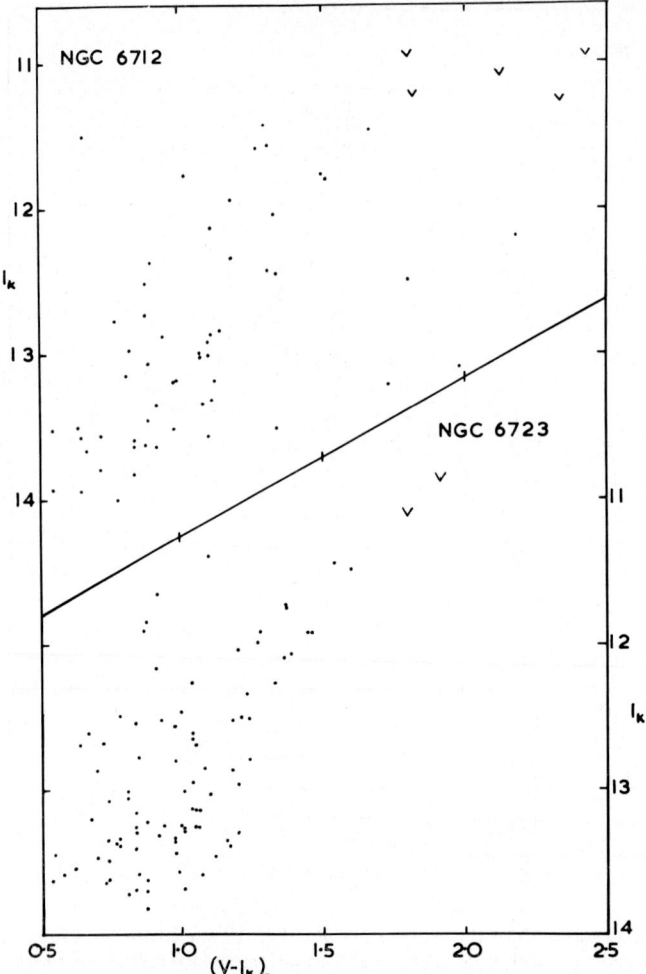

Fig. 2. *Top*: Stars in the region of NGC 6712, mainly those observed by Sandage and Smith (1966). *Below*: Stars within 4′ of the centre of NGC 6723. $(V-I_K)$ but not I corrected for reddening.

(6) There is no obvious clumping of variables in colour along the giant branch to parallel the clumping in period of variability reported by Arp *et al.* (1963).

The remaining clusters contain fewer measurable stars than NGC 104 and are perhaps best considered in groups suggested by the original selection criteria:

(A) $-0.35 \leqslant Q \leqslant -0.28$. Some blue horizontal branch and RR Lyrae stars. NGC 6171, 6712, 6723.

(B) $-0.26 \leqslant Q \leqslant -0.20$. Red horizontal branch, very few RR Lyrae stars. NGC 104, 6356, 6388, 6637.

(C) $-0.13 \leqslant Q \leqslant -0.09$. Red horizontal branch, very few RR Lyrae stars. NGC 5927, 6352, 6553.

These groups show the following properties.

(A) The amplitudes of variability are quite large even near $(V-I_K)=1.8$, where variable stars are first found. The clusters are too sparse to permit a decision as to the existence of a gap to the blue of $(V-I_K)=1.8$. The reddest variable has $(V-I_K)\sim 2.4$, which is uncertain inasmuch as NGC 6712 lies in a very crowded field where cluster membership requires confirmation from radial velocities.

(B) NGC 6356, where only the tip of the giant branch has been observed, shows variables between $(V-I_K)=2.0$ and 2.9, intermingled with 'non-variable' stars as in the case of NGC 104. NGC 6388 and NGC 6637, however, seem to show a sudden onset of variability with most of the stars beyond the limiting colour detected as variable on a moderate number of plates. Both these clusters as well as 47 Tuc contain variables (R10 in 47 Tuc, V3 in NGC 6637, V3 and V10 in NGC 6388) which appear brighter and in some cases bluer than the remaining variables. The spectral type of Radcliffe 10 in 47 Tuc (Feast and Thackeray 1960) is K3II, in agreement with the blue colour of $(V-I_K)=1.60$. The explanation for these stars may be in duplicity or in their being observed, by chance or perhaps as a result of a long wave in the light curve,

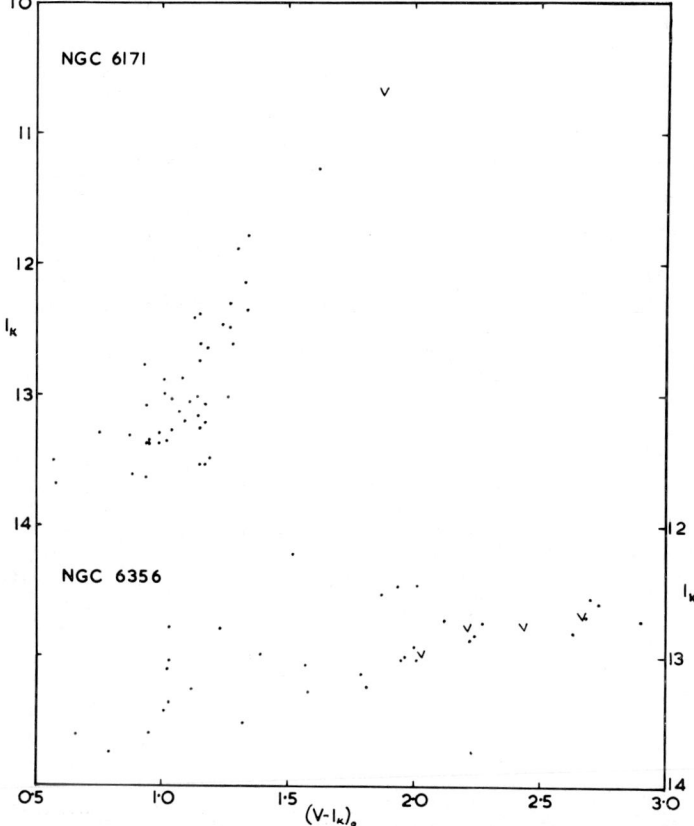

Fig. 3. *Top*: Stars in NGC 6171, mostly selected from those observed by Sandage and Katem (1964). *Below*: Stars in NGC 6356, mostly selected from those observed by Sandage and Wallerstein (1960). $(V-I_K)$ but not I has been corrected for reddening.

always near maximum; the few other photographs or magnitudes available from earlier work on 47 Tuc and NGC 6637 show the stars always bright. Other possible explanations are that they are at a brighter (and later?) evolutionary stage where instability sets in at a higher temperature or, speculatively, an enhanced C/O ratio prevents the formation of TiO despite a low temperature. Eggen (1972) found a few similarly displaced stars in metal poor clusters, for which the last suggestion seems inappropriate.

(C) NGC 5927 contains variables of considerable amplitude with $2.3 < (V-I_K) < 3.0$. The absence of bluer variables reflects the lack of stars with $1.9 < (V-I_K) < 2.3$; the reality of either effect is uncertain, but the distribution of stars with colour is clearly different from that in 47 Tuc, and the slope of the giant branch is less. NGC 6553 seems very similar, but NGC 6352, though too poor to contribute much to the discussion, has a giant branch whose slope seems rather greater.

The main conclusion from consideration of the different groups is that the variables

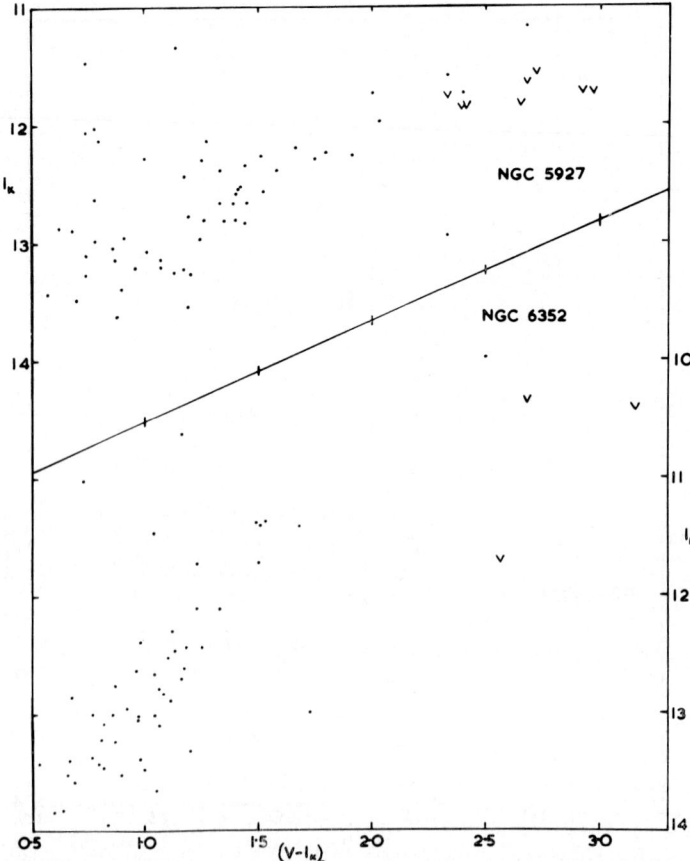

Fig. 4. *Top*: Stars within 2′ of the centre of NGC 5927. *Below*: Stars within 4′ of the centre of NGC 6352. $(V-I_K)$ but not I has been corrected for reddening.

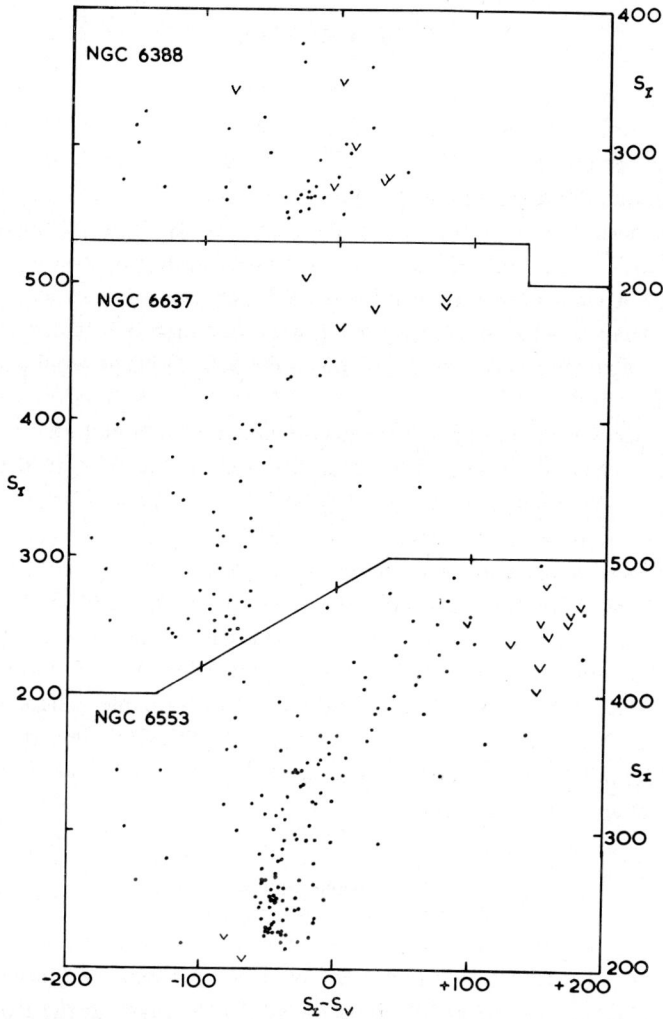

Fig. 5. Pseudo-colour-magnitude diagrams for *Top*: Stars within 3′.3 of the centre of NGC 6388. *Middle*: Stars within 2′ of the centre of NGC 6637. *Below*: Stars within 2′.2 of the centre of NGC 6553.

are redder on average as one goes from the metal-poor clusters to A and from A to C, which is generally thought to continue in the direction of increasing metal abundance. The observed shift could result both from a greater extension of the giant branch to low temperature and from higher metal abundance resulting in stronger TiO bands (the feature which has the largest effect on $(V-I_K)$) at given Te. We note, however, that the metal poor cluster ω Cen has, in addition to a group of variables with $(V-I_K)$ = 1.4, two (V6 and V17) with $(V-I_K) \sim 2.4$ (Dickens *et al.*, 1972). The situation may thus be more complex, though the finer details are only apparent in such rich clusters as ω Cen and 47 Tuc.

5. Mira Variables

Mira variables are readily detectable by their large amplitude and very red $(V-I_K)$ when faint. Several were recognised as such for the first time during this investigation. (Table I). The period of V10 in NGC 6637 has been found from Cape and Radcliffe plates to be about 195 days, similar to that of V4. V4 in NGC 6553 may have $P=270$ days, though periods near 150 and 210 days are perhaps not excluded. Osborne's (1968) variable, V1, in NGC 5927 has only twice been seen bright on Radcliffe plates, at $V\sim15$, on 1 August 1954 and on 1 April 1968. Negative observations, the variable being fainter than $V\sim18$ on 4 occasions and fainter than $V\sim16.5$ on 6 occasions, argue strongly against a period near 200 days. $P=300-330$ days would satisfy all the negative observations.

These long periods would normally be accompanied by a fainter M_V(max) than that for the 200 day variables, in line with the gentler slope of the giant branch in the clusters of group C. We might regard V42 ($P=149^d$) in ω Cen as a Mira variable on the strength of the spectroscopic results of Dickens et al. (1972), in which case there would be a trend of increasing period with increasing metal abundance, from 150^d in ω Cen to $\sim200^d$ in NGC 6712, 6637 and 104, and to $\sim300^d$ in NGC 6553 and NGC 5927. Confirmation of this result is highly desirable and determination of definitive periods for the eight outstanding cases (V1, V3, (V4) in NGC 6356; V1, V2, V4 in NGC 6388; V1 in NGC 5927; V4 in NGC 6553) would be most valuable. These variables are almost all severely crowded or located in regions of high surface density of faint stars, so that a moderately large telescope will be needed, preferably using infrared plates to suppress the bluer stars in the clusters.

Acknowledgements

We are indebted to Dr A. D. Thackeray for the loan of his plates and finding chart for NGC 6553: TLE is indebted to Dr M. W. Feast for many valuable discussions; to Mr G. A. Harding for the use of plates of NGC 6637 taken at the Royal Observatory, Cape; and to Professor D. W. N. Stibbs for his hospitality at the University Observatory, St. Andrews, where most of the blink microscope work was done. J. W. M. wishes to acknowledge the support of the Radcliffe-Henry Skynner Fellowship and of the Science Research Council during this project.

References

Arp, H., Brueckel, F., and Lourens, J. V. B.: 1963, *Astrophys. J.* **137**, 228.
Bergh, S. van den: 1967, *Astron. J.* **72**, 70.
Brooke, A. L.: 1969, Thesis, Australian National University.
Catchpole, R. M., Feast, M. W., and Menzies, J. W.: 1970, *Observatory* **90**, 63.
Dickens, R. J., Feast, M. W., and Lloyd Evans, T.: 1972, *Monthly Notices Roy. Astron. Soc.*, **159**, 337.
Eggen, O. J.: 1972, *Astrophys. J.* **172**, 639.
Feast, M. W.: 1972, *Quart. J. Roy. Astron. Soc.* **13**, 191.
Feast, M. W. and Thackeray, A. D.: 1960, *Monthly Notices Roy. Astron. Soc.* **120**, 463.

Fourcade, C. R. and Laborde, J. R.: 1966, *Atlas y Catálogo de Estrellas Variables en Cúmulos Globulares al sur de* $-29°$, Cordoba.
Hartwick, F. D. A. and Hesser, J. E.: 1972, *Astrophys. J.* **175**, 77.
Hartwick, F. D. A. and Sandage, A. R.: 1968, *Astrophys. J.* **153**, 715.
Hogg, H. S.: 1955, *Publ. David Dunlap Obs.* **2**, 35.
Kron, G. E. and Smith, J. L.: 1951, *Astrophys. J.* **113**, 324.
Lloyd Evans, T. and Menzies, J. W.: 1971, *Observatory* **91**, 35.
Osborne, W.: 1968, *Observatory* **88**, 26.
Rosino, L.: 1962, *Mem. Soc. Astr. Ital.* **33**, 351.
Sandage, A. R. and Katem, B.: 1964, *Astrophys. J.* **139**, 1088.
Sandage, A. and Smith, L. L.: 1966, *Astrophys. J.* **144**, 894.
Sandage, A. R. and Wallerstein, G.: 1960, *Astrophys. J.* **131**, 598.
Wildey, R. L.: 1961, *Astrophys. J.* **133**, 430.

DISCUSSION

Hogg: How many of these variables are new?
Lloyd-Evans: About 50.

NARROW-BAND PHOTOMETRY OF RED VARIABLES IN GLOBULAR CLUSTERS

ROBERT F. WING*

Perkins Observatory, The Ohio State and Ohio Wesleyan Universities, Ohio, U.S.A.

Abstract. Fourteen red variables in the southern globular clusters 47 Tuc, ω Cen, and NGC 362 have been observed on an eight-color system of narrow-band photometry in the near infrared. Temperatures are derived from blackbody fits to the calibrated fluxes, and spectral types are given for the M stars. The types observed for the three Mira variables in 47 Tuc range from M3.1 to M7.5; two small-range variables in the same cluster are later than M4. The variables in ω Cen are mostly earlier than K5, but spectra of types M3 and M0 were also encountered among radial-velocity members. In both the metal-rich 47 Tuc and the metal-poor ω Cen, the relation between TiO band strength and temperature is approximately normal. Several of these stars fall well above or below the red giant branches of their clusters in diagrams of infrared magnitude against temperature. Comparisons are made with recent results obtained at Radcliffe Observatory on some of the same stars.

1. Introduction

Red variable stars have long been known to occur in several globular clusters. Approximately one-fourth of the globular clusters listed in the recent catalogue by Sawyer Hogg (1973) contain variables which are cooler than Cepheids, or at least seem likely to be from the data given. Most of these are semi-regular variables with periods between 50^d and 150^d; others are irregular variables, or Miras with periods near 200^d.

Although many of these stars are among the brightest members of their clusters, rather little is known about their spectra. Despite the efforts of Joy (1949), Feast and Thackeray (1960), and others who have furnished spectral classifications based upon slit spectrograms, as well as the recent work at the Radcliffe Observatory described at this Colloquium by Feast (1973), it can still be said that the spectra of the majority of known red variables in globular clusters have never been observed. Indeed, the discovery of many new red variables in ten clusters reported here a few minutes ago by Lloyd Evans and Menzies (1973) shows how incomplete has been our knowledge of even the occurrence of such stars.

Narrow-band photometry often compares favorably with slit spectroscopy as a technique for classifying spectra. Normally there is a gain in speed and internal accuracy, but a loss of spectroscopic detail and an increased likelihood of misinterpretation. For most spectral types the features used spectroscopically as classification criteria are too weak to be measured conveniently by photometry; the photometrist must then measure some stronger feature which correlates with spectral type, and rely upon this correlation in assigning a type. In the case of the M stars, on the other hand, the primary spectroscopic criteria are the TiO bands, which can themselves be

---

* Visiting astronomer, Cerro Tololo Inter-American Observatory, which is operated by the Association of Universities for Research in Astronomy, Inc., under contract with the National Science Foundation.

measured easily and accurately with narrow-band filters. Experience has shown that a two-color TiO index measured photoelectrically with a telescope of moderate size (40–60 in.) can give a classification for an early M star of visual magnitude 12, to the accuracy achieved spectroscopically on bright stars (about one-quarter of a sub-type), in less than 10 min. In practice, I prefer to spend more time in order to measure at additional wavelengths, so as to establish the level of the continuum as well as possible in each star and to test the presence of other molecules in addition to TiO. On the eight-color system used here, the additional measurements yield color temperatures, infrared magnitudes, and band strengths of CN and VO.

The present study was undertaken to provide spectral types and color temperatures for a sample of cool cluster variables and also to test the effectiveness of the eight-color system on fainter stars than had previously been studied in this manner. Fourteen variables in 3 southern clusters (5 in 47 Tucanae, 8 in ω Centauri, and 1 in NGC 362) were observed either once or twice each in the course of three trips to Cerro Tololo Inter-American Observatory (CTIO) in 1970 and 1971. Similar observations in the northern hemisphere are planned but so far have been prevented by poor weather.

The more extensive observing program begun at Radcliffe Observatory in 1971 and discussed by Feast (1973) has been entirely independent of my work but has had essentially the same objectives. By augmenting image-tube spectrograms with infrared wide-band photometry, the Radcliffe observers have been able to obtain spectral types, temperature indices, and infrared magnitudes, i.e. the same information as provided by the eight-color photometry. It is not surprising that they too have given initial emphasis to the variables in the bright, rich clusters 47 Tuc and ω Cen. A comparison of our results is given in the last section of this paper.

2. Observations

The eight-color photometric system, designed in 1969 on the basis of earlier scanner measurements made at Lick Observatory, uses interference filters averaging 55 Å in width and ranging in wavelength from 7100 Å to 11000 Å. Three molecules are specifically measured by the program: TiO by filter 1 (the filters are numbered in order of increasing wavelength), CN by filters 4 and 8, and VO by filter 6. In late M stars the continuum is defined by the measurements in filters 2 and 5, while filters 3 and 7 serve this purpose in carbon stars. In K and early M stars, several of the filters measure nearly clean continuum regions. The measurements are reduced to a system of absolute fluxes so that comparisons with blackbodies or model atmospheres can be made. More detailed information on the photometric system is given in Wing (1971), where representative eight-color 'spectra' are also shown.

The selection of stars to observe was based largely upon the atlas of Fourcade *et al.* (1966), on which are marked all known variables in globular clusters south of declination $-29°$. Since most of the red variables in this atlas are found in the two giant clusters 47 Tuc and ω Cen, these clusters were given highest priority and, with one exception, the variables in other clusters have not yet been observed on this program.

A comparison of the red variables in these two clusters is of interest because 47 Tuc is much richer in metal content than ω Cen.

At maximum light the red variables in 47 Tuc and ω Cen are near $V=11$ or 12; at minimum they are typically about $V=12$ or 13, although the Mira variables in 47 Tuc reach magnitudes close to $V=15$. The $I(104)$ magnitudes of these stars, measured with filter 5 at 10400 Å, lie in the interval 7.7 to 9.7 on a scale on which Vega is zero. Approximately 20 min with the CTIO 60-in. telescope, including setting time and sky measurement, were spent on each eight-color observation. The accuracy of the fluxes can most realistically be estimated from the repeated observations that were made of several non-variable red giants in ω Cen, which will be reported elsewhere (Wing, 1973). The probable errors were found to be typically ± 0.05 mag., with photon statistics and uncertainties in the sky background contributing about equally. Contributions from other stars included within the diaphragm were usually negligible but in a few cases reached about 10%.

A. 47 TUCANAE (NGC 104)

This is a metal-rich cluster with few RR Lyrae stars and a strong condensation toward the center. Color-magnitude diagrams have been published by Wildey (1961) and Tifft (1963), while Eggen (1961, 1972) and Arp *et al.* (1963) have monitored several of the red variables. It has been known from the work of Feast and Thackeray (1960) that the giant branch defined by the non-variables extends to types as late as M2, and that several of the variables are also of type M2 or M3. The presence of three Mira variables long seemed a unique characteristic of 47 Tuc, but several Miras belonging to other metal-rich clusters have recently been found (see Feast, 1973).

Eight-color photometry has been obtained for 5 variables in 47 Tuc, namely the three Miras (V1, V2, and V3), the 165^d semi-regular variable V4, and the irregular variable V11. All five were observed in succession on 1970 June 30, within a time interval of 80 min; two of them, V1 and V4, were also observed on 1970 January 2. Prior knowledge of the spectra of these stars was rather meagre since Feast and Thackeray (1960) observed the three Miras only near maximum light and did not observe V4 or V11.

The eight-color spectra obtained on 1970 June 30 for the three Mira variables are plotted in Figure 1. As it happens, V1 was near maximum light, V2 was at an intermediate phase, and V3 was near minimum on this date. Their visual magnitudes, estimated crudely at the telescope, were 12.0, 12.5, and 14.5, respectively. By contrast, we see in the figure that their $I(104)$ magnitudes range over only about 0.7 mag., so that infrared observations are hardly more difficult at minimum than at maximum. This behavior is consistent with that of Miras in the field, for 25 of which Lockwood and Wing (1971) derived a mean amplitude of 1.0 mag. in $I(104)$. It is likely that these Miras are never as faint as $I(104)=9.0$, and they are probably the only members of any globular cluster capable of reaching magnitudes brighter than $I(104)=8.0$.

The procedure normally used to analyze spectra on the eight-color system begins with finding the blackbody curve that passes through one of the points in each group

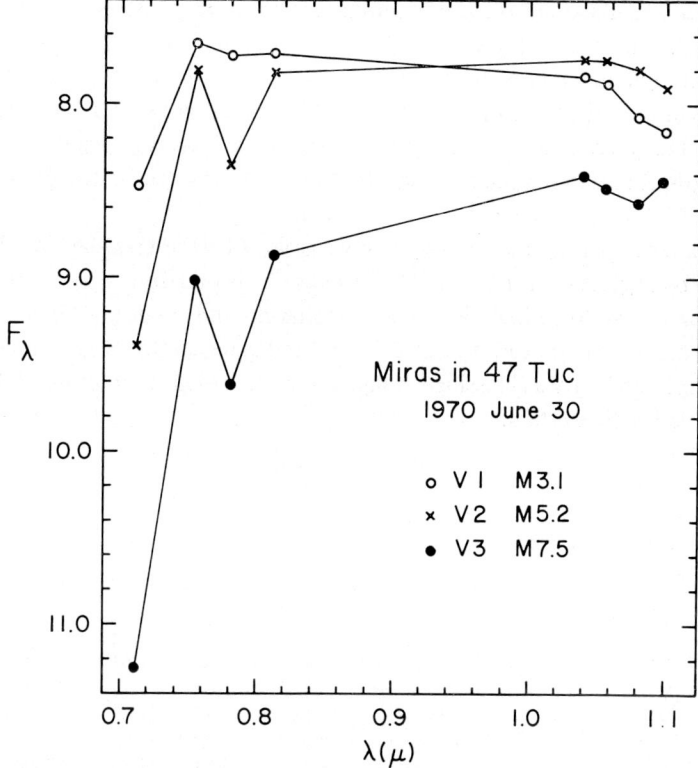

Fig. 1. Eight-color spectra of the three Mira variables in 47 Tuc obtained on 1970 June 30 at the CTIO 60-in. telescope. The fluxes are expressed on a magnitude scale. The TiO molecule is responsible for the depressions at filters 1 and 3.

of four filters and above the other points, giving directly the reciprocal color temperature $\theta = 5040/T$. Magnitude differences between the observed flux and the blackbody continuum flux at the same wavelength are then used as absorption indices, and are independent of reddening. The depression by TiO at filter 1 has been calibrated in terms of spectral type by Wing and Keenan (1973) for giant stars in the range K4 to M6. Types obtained in this manner have internal accuracies of ± 0.1 sub-type for bright stars, but in the present case of single observations of faint M stars the probable errors are about ± 0.2 sub-type.

In Figure 1, note that the spectra of V1 and V2 cross each other and that the depression of filter 1 in V3 exceeds 2 mag. V1 was clearly the warmest of the three Miras on this date and had the weakest TiO bands, while V3 was the coolest and latest. The types for V1 and V2, from the calibration of Wing and Keenan, are $M3.1 \pm 0.2$ and $M5.2 \pm 0.2$, respectively. The calibration of the TiO index does not extend beyond M6, since at later types the VO molecule appears and not only depresses the continuum at filter 2, thereby affecting the TiO index, but also supplies a more sensitive classification criterion at filter 6. Since V3 was decidedly later than M6 at the time of observation, it was classified on a preliminary scale for very late types that uses both TiO and VO. The resulting type, M7.5, has a maximum uncertainty of ±0.5 sub-type; the great strength of TiO and the weakness of VO show that the type cannot be earlier than M7.0 or later than M8.0, respectively. The depression at filter 6 amounts to only about 2 standard deviations; if it is real and due to VO, as is normal for stars of this TiO strength, it represents the first detection of the VO molecule in a globular cluster star.

Both observations of the semi-regular variable V4 were classified M4.0, but this type must be regarded as a lower limit because a second star, only about 1.5 mag. fainter visually, had to be included in the diaphragm and no doubt 'filled in' the TiO band to some extent. Indeed, the color of V4 indicates a later type (see below). The irregular variable V11 was classified M4.4. To my knowledge, no other classifications are available for these stars.

TABLE I

Red variables in 47 Tucanae

Star	Period	Date	$I(104)$	θ	θ_0	T_0	Spectral type
V1	212^d	'70 Jan. 2	7.72	1.63	1.58	3190 K	M4.0
		70 June 30	7.8	1.37	1.32	3820	M3.1
V2	203	70 June 30	7.7	1.58	1.53	3290	M5.2
V3	192	70 June 30	8.4	2.04	1.99	2530	M7.5
V4	165	70 Jan. 2	8.18	1.77	1.72	2930	> M4.0^a
		70 June 30	8.5	1.90	1.85	2720	> M4.0^a
V11	irr.	70 June 30	8.5	1.61	1.56	3230	M4.4

^{a} The spectral type of V4 is affected by a nearby star. The color indicates a type near M5.5

The results for stars in 47 Tuc are summarized in Table I. The star names and the periods are as given in Sawyer Hogg (1973). The next two columns give the dates of observation and the $I(104)$ magnitudes; the magnitudes obtained on 1970 June 30 are uncertain by ±0.1 mag. and are given to only one decimal place because of flexure within the cold box on that night. Next is given the reciprocal temperature θ obtained by fitting blackbody curves to the observed fluxes. These fluxes are affected by interstellar absorption which, however, is quite small in this case. Following Wing *et al.* (1973) the value $E(B-V) = 0.06$ will be adopted for 47 Tuc. For a normal reddening law, the effect of this amount of interstellar material is to make the observed θ's too large by 0.05, while the corresponding absorptions in V and $I(104)$ are $A(V) = 0.18$ and $A(104) = 0.06$ mag., respectively. The corrected inverse temperatures θ_0 and the corresponding corrected temperatures T_0 are listed in columns 6 and 7 of Table I, and the last column gives the spectral type.

It may be noted that the temperatures and spectral types given in Table I are only rather loosely correlated with one another. To examine this effect, the TiO indices in units of 0.01 mag. have been plotted against the derived values of θ_0 in Figure 2, and

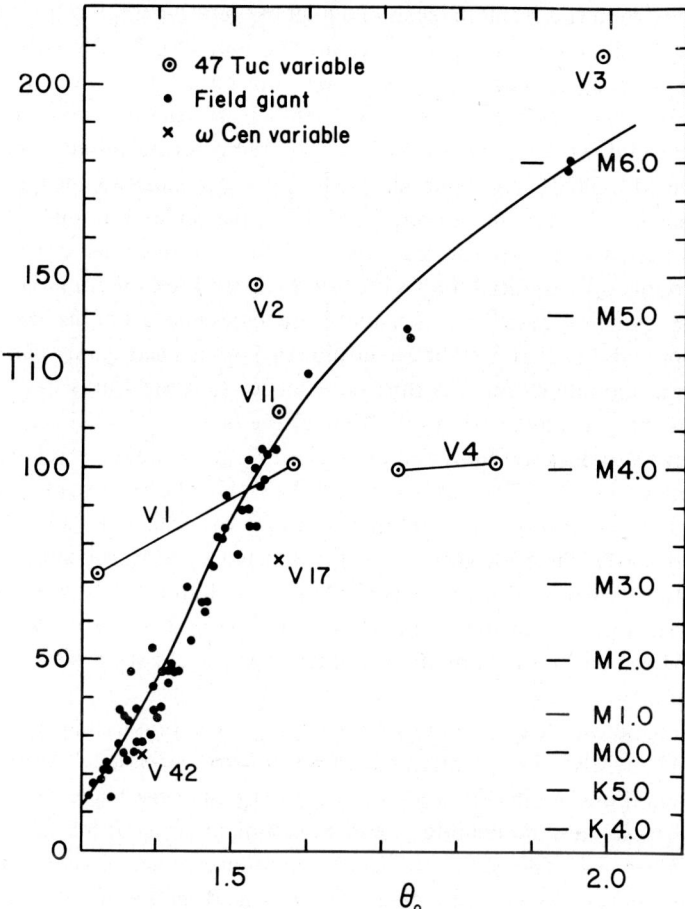

Fig. 2. The TiO index, defined as the depression at filter 1 in units of 0.01 mag., is plotted against the reciprocal color temperature θ, corrected for reddening as described in the text. The curve has been drawn through observations of bright, nearby giant stars. Different symbols are used to distinguish members of 47 Tuc and ω Cen. The calibration of the TiO index into spectral type is shown at the right.

the spectral-type calibration is shown near the right edge of the figure. For comparison with the 47 Tuc stars, a number of bright M-type giants (non-variables and small-range variables, assumed to be unreddened) observed with a CTIO 16-in. telescope in December 1969 are plotted as filled circles, and a smooth curve has been drawn through these points. For these bright stars the probable errors amount to one dot diameter in TiO and about two dot diameters in θ; for the variables in 47 Tuc the probable errors are between 2 and 3 times larger but are still much smaller than the scatter that is evident in the diagram.

Three of the four observations of Miras lie to the left of the mean giant relation in Figure 2, indicating that their colors are too blue for their types, or equivalently that their TiO bands are abnormally strong for their temperatures. Although one might be tempted to interpret this result in terms of a high oxygen-to-carbon ratio in the 47 Tuc stars, it is probably more to the point to recall that normal field Miras show the same effect. Some years ago I found that Miras trace out large loops in diagrams of band strength against color such as Figure 2, and that they are usually bluer than non-Miras of the same band strength (Wing, 1967b). The temperature found here for V1 at spectral type M3, 3820 K, corresponds to that of a normal giant of type K5. Such discrepancies occur commonly among field Miras, particularly among those of relatively short period and early spectral type, and the well-documented case of R Tri has been described by Spinrad and Wing (1969). Lockwood (1972) has explored the effect further, and his Figure 7 shows that at least three-quarters of his observations of Miras with periods less than 300^d lie on the blue side of the mean giant relation between band strength and color. There is thus no evidence that the Miras in 47 Tuc differ either in O/C or in behavior from the Miras of the field.

Small-range variables normally fall quite close to the mean giant relation between band strength and color. This is the case with the single observation of V11, but both observations of V4 show it to be abnormally red for its measured TiO band strength. As we noted above, the photometry of V4 is affected by a nearby star, the effect of which should be greatest at filter 1 where M stars are faintest. Using the mean value of θ_0 obtained for V4 and the mean relation of Figure 2, we estimate the spectral type to be M5.5. This result should be confirmed with a slit spectrogram that avoids the companion.

The (0, 0) band of CN measured by filter 8 can be seen in normal giants with types as late as M3 or M4, and it is often enhanced in Mira variables. Unfortunately the noise in observations of faint stars is greater at filter 8 than elsewhere because of lower detector sensitivity, and no definite statement can be made as to the CN strengths of the red variables in 47 Tuc. If the data are taken at face value, however, they indicate a somewhat weaker CN strength than in typical field gaints of the same types, in accord with the finding of Feast and Thackeray (1960) for the K giants of the cluster. The integrated light of 47 Tuc has likewise been found by Wing *et al.* (1973) to have weaker CN absorption than is found in normal solar-neighborhood giants.

V11 is identical to star 12 in Wildey's (1961) paper. Wildey suggested that it may be a field star because it falls away from the main giant branch, being both redder in $B-V$ and somewhat fainter visually than the stars at the tip. Its measured color, $B-V=1.91$, is hard to account for, whether or not it is a member, in view of the small reddening in this direction, and it should be checked photoelectrically. However, its depressed V magnitude can readily be understood in terms of the TiO absorption that occurs at this spectral type, so that the photometry seems rather to support its membership in the cluster.

It will be useful to determine the mean absolute $I(104)$ magnitude of the Mira variables in 47 Tuc for eventual comparison with Miras in other clusters and in the field. At present the best we can do is adopt $I(104)=8.0$ for the mean apparent magnitude of these stars; with the true distance modulus $(m-M)_0=13.49$ given by Arp (1965) and the absorption $A(104)=0.06$ quoted above, we then obtain $M(104)=-5.6$

for the absolute magnitude at mean light. Clearly many more observations should be made to improve the determination of the apparent magnitudes at mean light and at typical maxima.

It is intended to publish the eight-color fluxes from the observations discussed here together with additional observations in 47 Tuc planned for a forthcoming run at Cerro Tololo. Observations of non-variable red giants as well as variables are planned.

B. ω CENTAURI (NGC 5139)

In contrast to 47 Tuc, ω Cen is metal-poor, contains well over 100 RR Lyrae stars, and has a relatively open structure so that measurements of individual stars in the central regions are much less affected by crowding. It also differs from 47 Tuc in having a very large radial velocity, making it possible to test the membership of any star quite definitely once a slit spectrogram is available. Color-magnitude diagrams have been published by Woolley (1966) and Geyer (1967).

Prior to 1970 no M-type stars were known to occur in this cluster apart from the Mira variable V2, which Feast (1965) has shown to have a grossly discordant radial velocity. On the other hand, ω Cen is the only globular cluster known to contain carbon stars; three of these have now been found (Harding, 1962; Stock and Wing, 1972; Dickens, 1972), and two of them have been confirmed as cluster members by radial velocity measurements. Although 11 slow variables have been found which are known from the photometry of Woolley (1966) to be redder than $B-V=1.30$, their spectra were largely unknown prior to this study and the work at Radcliffe Observatory (Dickens *et al.*, 1972; Feast, 1972, 1973). Several M-type spectra have now been found

TABLE II

Red variables in ω Centauri and NGC 362

Star	Period	Member-ship	Date	$I(104)$	θ	θ_0	T_0	Spectral type
ω Centauri								
V2	236^d	NM	'70 June 30	7.81:	1.45	1.33	3790 K	M2.8
			71 June 14	8.34	2.81	2.69	1870	M8.0:
V17	65	M	71 June 14	9.37	1.68	1.56	3230	M3.2
V42	149	M	71 June 18	8.52	1.50	1.38	3650	M0.0
V53	33 or 70	M	70 June 30	9.70:	1.50:	1.38:	3650:	<K5
			71 June 16	9.76	1.54:	1.42:	3550:	<K5
V138	75	?	70 July 1	9.32:	1.32:	1.20:	4200:	<K5
			71 June 18	9.23	1.37	1.25	4030	<K4
V148	90:	?	71 June 15	9.12	1.30	1.18	4270	≤K4.5 (str. CN?)
V152	irr.	M	70 July 1	8.68:	1.36	1.24	4060	<K5
			71 June 18	9.04	1.42	1.30	3880	≤K4
V164	irr.	M	70 July 1	9.39:	1.51	1.39	3630	<K5
			71 June 14	9.70	1.49	1.37	3680	≤K4 (str. CN?)
NGC 362								
V2	90	?	71 June 15	10.32	1.19	1.19	4240	≤K4.5

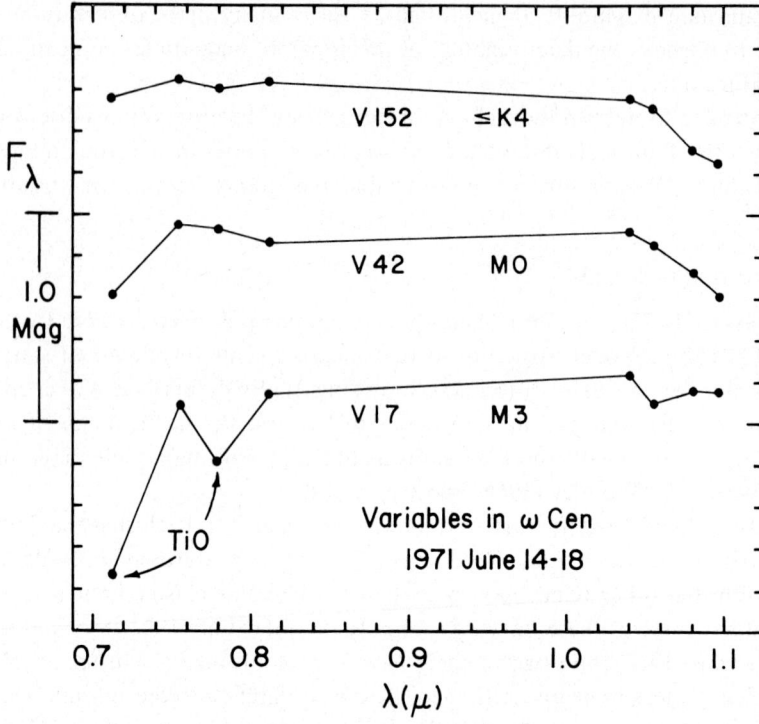

Fig. 3. Eight-color spectra of three small-range red variables in ω Cen. Different zero points are used for each star in this presentation; the actual $I(104)$ magnitudes are given in Table II.

among these variables. *UBVRI* photometry for some of them has recently been published by Eggen (1972).

Eight-color photometry was obtained in 1970 and 1971, again with the CTIO 60-in. telescope, for the Mira variable V2 and seven other red variables. Six of the latter were also observed by Dickens *et al.* (1972), and a comparison of our results is given in the concluding section of this paper. The observations are listed in Table II, which has the same arrangement as Table I except that a column has been added to indicate membership (M for member, NM for non-member) according to radial-velocity measurements (Feast, 1965; Dickens *et al.*, 1972). The $I(104)$ magnitudes measured on 1970 June 30 and July 1 are uncertain by about ±0.05 mag. because of the flexure problem mentioned earlier and are therefore given with colons; the uncertainties are smaller here than in 47 Tuc because local standards were observed. The adopted interstellar reddening, following Wing and Stock (1973), is $E(B-V)=0.15$, corresponding to $\Delta\theta=0.12$, $A(V)=0.45$, and $A(104)=0.15$ mag.

Figure 3 shows the eight-color spectra of three of the red variables in ω Cen. V17 has strong TiO bands corresponding to type M3.2±0.2. The small apparent depression at filter 6 in this star is probably just the result of noise in the data, since the VO band does not normally appear until the TiO is much stronger. V42 also shows definite

TiO absorption and is assigned type M0.0±0.3, which means a range of possible values from K5.7 to M0.3 since types K6–K9 do not exist on the scale employed.

V152 is shown as an example of a star having very weak or absent TiO bands. If the slight depression at filter 1 is real and due to TiO, the type is K4.0. It is also possible, however, that no TiO is present and the depression at filter 1 is due merely to a combination of observational scatter and weak CN absorption, in which case the type is earlier than K4.

The variables V53 and V138 show no evidence of TiO absorption, and only limits can be placed on their types. In the case of stars without TiO, earlier limits correspond to more accurate observations. In general, the 1971 data are of superior quality to the 1970 data.

The spectra of V148 and V164 have weak depressions at filter 1 that are believed to be real. If they are, there are two possible interpretations: (1) TiO is present, in amounts corresponding to types K4.5 and K4.0, respectively; or (2) TiO is absent and CN is abnormally strong, affecting filter 1. This second possibility is suggested by the depressions at filters 4 and 8, which are intended to measure CN. Choosing between these alternatives is simply a matter of securing further observations to reduce the observational scatter so that the level of CN absorption can be established from the measurements at filters 4 and 8, since then the contribution of CN to filter 1 would be known (see White, 1972). In view of the presence of carbon stars with much stronger CN bands in this cluster, it would be particularly interesting to confirm that stars with more modest CN enhancements also occur.

The two small-range variables showing definite TiO absorption, V17 and V42, have been plotted as crosses in Figure 2. We see that they have nearly normal band strengths for their colors, after corrections for reddening. Conversely, if we use Figure 2 to determine the reddening of ω Cen by assuming a normal relation between band strength and color (which may or may not be appropriate), the indicated reddening in $B-V$ is in the range 0.15 to 0.25 mag.

We may use the $I(104)$ magnitudes and the values of θ given in Table II to construct an infrared color-magnitude diagram that is essentially free of the effects of blanketing. This diagram will be shown in a subsequent paper (Wing, 1973) which will also include the complete eight-color photometry for stars observed in ω Cen, including 12 non-variable red giants defining the giant branch. Here we simply call attention to the fact that these variables scatter widely about the giant branch, even in a diagram that should closely approximate one of bolometric magnitude against effective temperature. At the time of these observations, the M3 star V17 was much cooler than any of the non-variables, while V42 was some 0.8 mag brighter than the tip of the giant branch. Of the non-TiO stars, V53 and V164 fall below the tip, while V138, V148, and V152 are brighter and warmer than the non-variables. The general locations of these stars are confirmed by the V, I photometry of Dickens *et al.* (1972), who also verified the cluster membership of most of them. Lloyd Evans and Menzies (1973) have also called attention to the existence, in other clusters, of stars that are brighter and bluer than the stars of the giant branch.

Finally, mention should be made of the Mira variable V2. Although Feast (1965) classified it as a 'normal Me variable', the present observations are the first, to my knowledge, to give spectral sub-types. The more advanced of the two types, M8.0:, was measured at $V \approx 16$ and is uncertain because it was necessary to include within the diaphragm a star that was brighter visually and that affected the TiO measurement at filter 1, where V2 was faintest; the classification is therefore based primarily on the VO strength at filter 6.

Since both the spectral range and the period (236^d) of V2 in ω Cen are rather similar to those of the Miras in 47 Tuc, it is surprising that its radial velocity (Feast, 1965) differs by more than 250 km s^{-1} from that of the cluster. Assuming $(m-M)_0 = 13.7$ (Eggen, 1972) and $A(104) = 0.15$ (Wing and Stock, 1973) for ω Cen, we find that if V2 lies at the same distance, then its infrared absolute magnitude, corresponding to the mean of the two observations, is $M(104) = -5.8$, in close agreement with the value -5.6 found above for the Miras in 47 Tuc. Therefore it seems likely that V2 does, in fact, lie at roughly the same distance as ω Cen, whether or not it is (or ever was) physically associated with it. Since the original velocity measurement was based on a single plate, it would be worthwhile to secure a second plate and repeat the measurement, and Dr Feast has indicated to me that he will make this observation at the next favorable opportunity.

C. NGC 362

A single observation of the 90^d variable V2, the only known slow variable in NGC 362, is listed in Table II. Although this cluster is quite small and compact, V2 is located far enough out from the center that crowding was not a problem. No definite TiO absorption was detected; the spectral type is not later than K4.5. The color temperature of 4240 K, assuming no reddening, suggests an early K spectral type.

Eggen (1972) has published *UBVRI* photometry for this star and several nonvariables belonging to NGC 362. V2 lies at the tip of the giant branch, and its $R-I$ colors are consistent with a spectral range of roughly K2 to K4.5.

3. Summary and Final Remarks

The eight-color narrow-band photometry has proved to be a useful tool in studying the red variables in at least the brighter of the globular clusters. Although it would be worthwhile spending more time on each star to increase the accuracy of the data, the present observations, representing only about 20 min per star, have yielded (1) more accurate spectral types for the M stars than have previously been available, (2) CN indices which readily distinguish carbon stars and indicate that a range in CN strength probably exists among the K giants of ω Cen, (3) color temperatures that are unaffected by blanketing, and (4) an $I(104)$ magnitude measured in the infrared continuum, near the peak of the energy distributions of late K and early M stars. The chief reasons for wishing to improve the observational accuracy are to locate these stars more precisely in the infrared color-magnitude diagram and to make more subtle distinctions with regard to CN strength.

As Feast (1973) has mentioned, the Radcliffe image-tube spectrograms of globular cluster stars are being supplemented by wide-band infrared photometry in I and K. When the spectroscopic and photometric observations are combined, the Radcliffe data give essentially the same information as the eight-color system. The spectrograms, of course, give more spectroscopic information than the narrow-band photometry; in particular they furnish a radial velocity. The eight-color system, on the other hand, is capable of measuring weaker CN bands and is more convenient for an observer working alone, since all the data are obtained simultaneously.

Despite the many differences between 47 Tuc and ω Cen, they have been found to contain rather similar small-range variables of type M. No systematic difference in the relation between TiO strength and temperature (Figure 2) between the two clusters is evident. A meaningful comparison of the CN strengths of the K and M stars must await more precise observations. One difference between the clusters that the data of Tables I and II seem to indicate is in the spectral types of the non-Mira variables: both such stars observed in 47 Tuc are later than any of their counterparts in ω Cen. However, exactly the opposite conclusion was reached by Feast (1973) who found no types later than M3 in 47 Tuc whereas Dickens *et al.* (1972) classified two stars in ω Cen as M5. The differences are simply the result of not observing all of the same stars: the Radcliffe program did not include the two latest small-range variables in 47 Tuc, just as I missed the two latest stars in ω Cen. When all the data are considered together, we must conclude that the coolest non-Miras in the two clusters are quite similar, with spectral types near M5. For the stars observed in common, the eight-color and Radcliffe types are in excellent agreement. The type given here for V42 in ω Cen, M0.0, lies outside the range M1:e–M2.5e recorded by Dickens *et al.* (1972) but may easily be attributed to the variability of the star. Likewise the range in type found here for the Miras in 47 Tuc, M3.1–M7.5, is consistent with the types M2–3 given for all three by Feast and Thackeray (1960), since the earlier spectroscopic observations were restricted to phases near maximum light.

One lesson learned from the infrared photometry – both wide-band and narrow-band – is that stars which fall below the tip of the red giant branch in conventional color-magnitude diagrams (V vs $B-V$) should not be dismissed as foreground dwarfs on that basis alone. Some years ago, I pointed out (Wing, 1967a) that the extension of a cluster's red giant branch into the M types, if it exists, should be found directly below the tip since the appearance of TiO depresses the V magnitude while holding the $B-V$ color constant throughout the range K5–M8. Several examples of such stars have recently been found: for example, V11 in 47 Tuc and V17 in ω Cen, considered non-members by Wildey (1961) and Eggen (1972), respectively, primarily on the basis of their faint V magnitudes, have both been found to have infrared magnitudes consistent with membership, and the latter star's membership is confirmed by its radial velocity (Dickens *et al.*, 1972). Several similar stars have been found by Lloyd Evans and Menzies (1973).

At the same time, there exist non-TiO variables which lie both above and below the giant branches of their clusters; since blanketing is not important in these stars, they have the same location relative to the giant branch no matter what kind of photometry is used. Good examples are R10 in 47 Tuc (above giant branch: Lloyd Evans and Menzies, 1973) and V164 in ω Cen (below giant branch: Eggen, 1972; Dickens *et al.*, 1972; Feast, 1973; and this paper). It will be an important theoretical problem to interpret the abnormal luminosities and variability of these stars.

It is to be hoped that spectroscopic work on very red stars in globular clusters will progress more rapidly now that several studies have shown them to be quite common. Direct photography in the infrared, such as Lloyd Evans and Menzies (1973) have described, is an efficient means of selecting stars for spectroscopic examination and should be applied to all globulars. A specific search for carbon and S-type stars in globular clusters should be made, and several techniques that might be used have been described by Wing and Stock (1973). Whereas the atomic spectra of heavy elements indicate the metallicity of the material from which the cluster formed, the molecular spectra of the red stars provide information about the relative abundances of carbon, nitrogen, and oxygen, and hence about the processes of nucleosynthesis and mixing that occur during the red giant stage of an individual star's life history.

Acknowledgements

It is a pleasure to acknowledge stimulating and useful discussions with Drs R. J. Dickens, M. W. Feast, T. Lloyd Evans, and J. W. Menzies, and to thank them for providing preprints of their work while this paper was being prepared. I am grateful to the Director of Cerro Tololo Inter-American Observatory, Dr V. M. Blanco, for the time on the 60-in. telescope that made these observations possible, and to the National Science Foundation for financial support.

References

Arp, H. C.: 1965, in A. Blaauw and M. Schmidt (eds.), *Galactic Structure*, University of Chicago Press, Chicago, p. 401.
Arp, H. C., Brueckel, F., and Lourens, J. v. B.: 1963, *Astrophys. J.* **137**, 228.
Dickens, R. J.: 1972, *Monthly Notices Roy. Astron. Soc.*, **159**, 7p..
Dickens, R. J., Feast, M. W., and Lloyd Evans, T.: 1972, *Monthly Notices Roy. Astron. Soc.*, **159**, 337.
Eggen, O. J.: 1961, *Roy. Observ. Bull.*, No. 29.
Eggen, O. J.: 1972, *Astrophys. J.* **172**, 639.
Feast, M. W.: 1965, *Observatory* **85**, 16.
Feast, M. W.: 1972, *Quart. J. Roy. Astron. Soc.* **13**, 191.
Feast, M. W.: 1973, in J. D. Fernie (ed.), *Variable Stars in Globular Clusters and in Related Systems*, D. Reidel Pub. Co., Dordrecht, Holland, p. 131.
Feast, M. W. and Thackeray, A. D.: 1960, *Monthly Notices Roy. Astron. Soc.* **120**, 463.
Fourcade, C. R., Laborde, J. R., and Albarracin, J.: 1966, *Atlas y Catálogo de Estrellas Variables en Cúmulos Globulares al Sur de* $-29°$, Córdoba.
Geyer, E. H.: 1967, *Z. Astrophys.* **66**, 16.
Harding, G. A.: 1962, *Observatory* **82**, 205.
Joy, A. H.: 1949, *Astrophys. J.* **110**, 105.

Lloyd Evans, T. and Menzies, J. W.: 1973, in J. D. Fernie (ed.), *Variable Stars in Globular Clusters and in Related Systems*, D. Reidel Pub. Co., Dordrecht, Holland, p. 151.
Lockwood, G. W.: 1972, *Astrophys. J. Suppl.* **24**, No. 209, 375.
Lockwood, G. W. and Wing, R. F.: 1971, *Astrophys. J.* **169**, 63.
Sawyer Hogg, H. B.: 1973, *Publ. David Dunlap Obs.* **3**, No.6, in press.
Spinrad, H. and Wing, R. F.: 1969, *Ann. Rev. Astron. Astrophys.* **7**, 249.
Stock, J. and Wing, R. F.: 1972, *Bull. Am. Astron. Soc.* **4**, 324.
Tifft, W. G.: 1963, *Monthly Notices Roy. Astron. Soc.* **126**, 209.
White, N. M.: 1972, in M. Hack (ed.), *Colloquium on Supergiant Stars*, Trieste, p. 160.
Wildey, R. L.: 1961, *Astrophys. J.* **133**, 430.
Wing, R. F.: 1967a, in M. Hack (ed.), *Colloquium on Late-Type Stars*, Trieste, p. 205.
Wing, R. F.: 1967b, Doctoral dissertation, University of California, Berkeley.
Wing, R. F.: 1971, in G. W. Lockwood and H. M. Dyck (eds.), *Proceedings of the Conference on Late-Type Stars*, Kitt Peak National Observatory Contribution No. 554, Tuscon, p. 145.
Wing, R. F.: 1973, in preparation.
Wing, R. F. and Keenan, P. C.: 1973, in preparation.
Wing, R. F. and Stock, J.: 1973, in press.
Wing, R. F., Warner, J. W., and Smith, M. G.: 1973, *Astrophys. J.*, **179**, 135.
Woolley, R. v. d. R.: 1966, *Roy. Observ. Ann.* No. 2.

DISCUSSION

Buscombe: What angular aperture of diaphragm was used in the integrated narrow-band photometry of 47 Tuc?

Wing: Circular diaphragms of 43″ and 68″ were used for 47 Tuc: thus only the central regions were included.

Feast: (1) I should make clear that the spectral types I gave for the Miras in 47 Tuc refer to maximum light – they get to much later types when below maximum.
(2) In view of your work I checked the velocity of V 2 ω Cen. It is definitely quite different from the cluster.

Walborn: The 'spectral types' to a tenth of a subclass may be telling us more about the observational accuracy of the photoelectric technique than about the classification accuracy, in view of possible additional contributors to the band strengths, such as abundance differences, for instance.

Wing: For bright stars, the photoelectric types are reproducible to a tenth of a subclass, but the intrinsic scatter in the TiO $-\theta$ relation shows that the types are not pure temperature indicators, as you say. The same is true, however, of the MK types of M giants, since they likewise are essentially based on absolute TiO band strength.

THE VARIABLE STARS IN NGC 6723

JOHN MENZIES

University Observatory, Oxford, England

Abstract*. A study has been made of the variables in the Southern Hemisphere globular cluster NGC 6723, which is suspected of being relatively metal-rich on the basis of its colour-magnitude diagram and of its integrated spectral type of G3. Seven new RR Lyrae stars and two bright red, probably semiregular, variables have been found and the suspected variable of Fourcade and Laborde has been confirmed. The complement of RR Lyrae stars is now 27, consisting of 4 c-type and 23 ab-type variables, the mean periods being $P_c = 0\overset{d}{.}292$ and $P_{ab} = 0\overset{d}{.}537$. On the basis of the two-colour diagram of the horizontal branch stars the cluster is considered to be virtually unreddened. Applying Christy's models to the data from this study we find the following parameters for the variables: $M_V = 1\overset{m}{.}10$, mass $= 0.42\ M_\odot$, and $Y = 0.4$.

DISCUSSION

Cox: In your diagram of the horizontal branch, I notice that the colour range of the c-type variables is not much occupied either by variable or non-variable stars. Can you comment?

Also are there other globular clusters which show such a dearth of c-type variables whereas the ab-type stars are of uniform abundance relative to the non-variable stars on both sides of the gap?

Menzies: The colour-magnitude diagram I showed is for a complete sample of stars in an annulus of width $2\overset{\prime}{.}4$ centered on the cluster. I have good colours for only 12 of the 21 variables in this region, but assuming the remainder to behave like these 12, I think the absence is real.

I don't know of any other cluster like this, but only relatively few have had their RR Lyrae stars and their colour magnitude diagrams studied in sufficient detail.

Schwarzschild: Are any efforts being made to work with larger plate scales to improve the quality of the photometric measurements?

Dickens: We have a series of plates of clusters taken at the Cassegrain focus of the Isaac Newton telescope (plate scale 6″ per mm) to investigate this question. One difficulty arises in the measurement of the large, soft images with conventional iris photometry and microphotometry of the images, which is very laborious, but is probably necessary to make full use of the information contained in the stellar images.

Schwarzschild: May I bring up a question of observational technique, not specifically directed to Dr Menzies but to all who have shown us new photographic photometric results? Crowding of star-images has frequently been mentioned during these last two days as a basic difficulty for accurate photometry in globular clusters. My question is: Are we nowadays adjusting our plate scale – by simple enlarging optics – to maximize our photometric accuracy under the crowded conditions or are we still taking most of our photometric plates (i.e. not the plates for discovery of variables by blinking) with the scales as they happen to come with the available instruments, without enlarging optics? If a really good plate has stellar images with half-power diameters of, say, one second of arc, the photometric information least troubled by background effects or close neighbours may be contained in a circle of one half second diameter. On the other hand the commonly used high-sensitivity plates give reasonable signal to noise ratios only for areas, say, 0.05 mm in diameter (strong exposures – for which we astronomers seem to be addicted to – do not generally improve the signal to noise ratio, but do waste telescope time). Under these conditions an optimum plate scale would seem to be ten sec of arc per millimeter. Are such optimized plate scales nowadays used?

* Details of this work will be published elsewhere.

Menzies: For my part, I have put up with the plate scale provided. Increasing the plate scale won't reduce crowding very much because of the effects of seeing.

Racine: This is a very important point. Although increasing the focal length does not reduce crowding (because of seeing limitations), it increases the capacity per image element of the receiver and hence allows the sampling of a smaller image element to achieve the same statistical accuracy. As to what astronomers are in fact doing... well... I must say that most still put up with the plate scale given to them. However there has been some trend toward the use of finer grain emulsion (IIIaJ) which also provides a capacity increase.

THE PERIOD-LUMINOSITY RELATION FOR CEPHEIDS IN GLOBULAR CLUSTERS

B. V. KUKARKIN and A. S. RASTORGOUEV

Sternberg Astronomical Institute, U.S.S.R.

The period-luminosity relation for Cepheids in globular clusters has been investigated many times (e.g. Fernie, 1964; Kwee, 1968; Frolov, 1970; Demers, 1971).

The method of determination of the apparent distance moduli was recently revised by Kukarkin and Russev (1972). Instead of using a single absolute magnitude for RR Lyrae variables, the magnitudes according to pulsation theory (Christy, 1966, 1971) were adopted. The inhomogeneity of the absolute magnitudes of the RR Lyrae variables had already been established long ago (Pavlovskaya, 1953), but it attracted attention only recently. The different methods for determining the distance moduli of globular clusters were calibrated according to the new absolute magnitudes of the RR Lyrae variables. The problem consisted in the determination of the absolute magnitudes of the Cepheids in globular clusters according to the apparent distance moduli.

The photographic observations of 9 Cepheids in the globular clusters M5, M10, M12, M13 and M80 were made by A. S. Rastorgouev in the B photometric system on

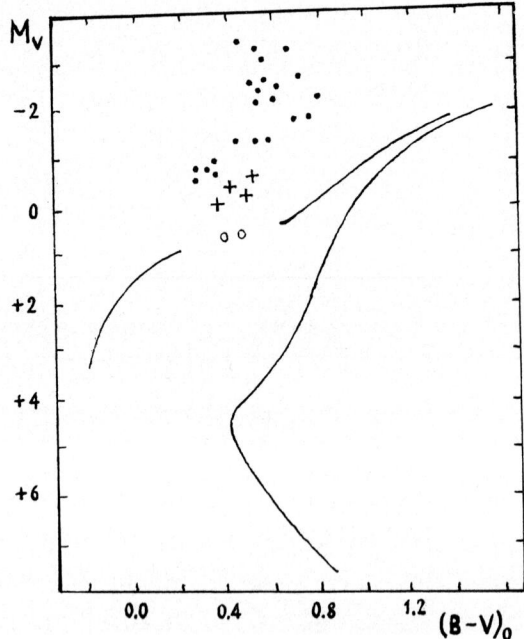

Fig. 1. Location of globular cluster Cepheids in the colour-magnitude away. Crosses represent peculiar Cepheids.

plates obtained at our Crimean Station. The results are given in Table I. (The observations will be published separately).

All the observations of Cepheids in globular clusters by different authors were reduced to the BV photometric system. The results are given in Table II. In the first column the NGC number of the globular cluster is given; in the second the number of

TABLE I
Crimean photographic observations

NGC	Var	log P	B
6205	V1	0.164	15.10
6205	V6	0.325	14.40
6205	V2	0.708	13.20
6254	V3	0.896	13.80
6218	V1	1.190	12.72
6093	V1	1.212	14.10
6254	V2	1.273	12.65
5904	V42	1.411	11.90
5904	V 84	1.423	12.05

TABLE II
Reduced photographic observations

NGC	Var	log P	$\langle V \rangle$	$\langle B \rangle$	$\langle B-V \rangle$	$\langle B-V \rangle_0$	mod$_V$	M_V	M_B
5139	43	0.063	13^{m}38	13^{m}84	0^{m}46	0^{m}32	13^{m}84	−0^{m}46	−0^{m}14
5139	92	0.129	13.96	14.45	0.49	0.35	13.84	+0.12	+0.47
5139	60	0.130	13.49	13.87	0.38	0.24	13.84	−0.35	−0.11
7078	1	0.158	14.85	15.17	0.34	0.23	15.30	−0.47	−0.24
6205	1	0.164	14.19	14.49	0.41	0.38	13.93	−0.15	+0.23
6402	76	0.276	15.84	16.60	0.76	0.27	16.33	−0.49	−0.22
6205	6	0.325	13.89	14.39	0.50	0.47	13.93	−0.04	+0.43
5139	61	0.357	13.44	14.07	0.63	0.49	13.84	−0.40	+0.09
6402	2	0.445	15.64	16.44	0.80	0.31	16.33	−0.69	−0.38
5139	48	0.651	12.77	13.41	0.64	0.50	13.84	−1.07	−0.57
6205	2	0.708	12.86	13.28	0.42	0.39	13.93	−1.07	−0.68
6254	3	0.896	12.80	13.59	0.79	0.56	13.92	−1.12	−0.56
6402	17	1.083	14.79	15.93	1.14	0.65	16.33	−1.54	−0.89
6402	7	1.133	14.75	15.97	1.22	0.73	16.33	−1.58	−0.85
5139	29	1.168	11.88	12.83	0.95	0.81	13.84	−1.96	−1.15
5272	154	1.185	12.48	12.97	0.49	0.47	14.82	−2.34	−1.87
6218	1	1.190	–	12.72	–	–	14.09	–	−1.54
7089	1	1.192	13.42	13.99	0.57	0.50	15.44	−2.02	−1.52
6093	1	1.212	13.44	14.20	0.76	0.58	15.45:	−2.01:	−1.43:
7089	5	1.244	13.29	13.86	0.57	0.50	15.44	−2.15	−1.65
6402	1	1.272	14.09	15.31	1.22	0.73	16.33	−2.24	−1.51
6254	2	1.273	11.74	12.56	0.82	0.59	13.92	−2.18	−1.59
7089	6	1.284	13.20	13.70	0.60	0.53	15.44	−2.34	−1.81
5904	42	1.411	11.28	11.84	0.56	0.51	14.33	−3.05	−2.54
5904	84	1.423	11.42	12.01	0.59	0.54	14.33	−2.91	−2.37
5139	1	1.465	10.89	11.68	0.79	0.65	13.84	−2.95	−2.30
7089	11	1.525	12.21	12.72	0.51	0.44	15.44	−3.23	−2.79

the variable according to the catalogue of Sawyer Hogg (1955); in the third the logarithm of the period; in the fourth and fifth the apparent mean magnitudes $\langle V \rangle$ and $\langle B \rangle$; in the sixth the value $\langle B-V \rangle$; in the seventh $\langle B-V \rangle_0$; in the eighth the apparent distance modulus mod_V; in the ninth and tenth the absolute magnitudes M_V and M_B.

In Figure 1 the positions of the Cepheids on the colour-magnitude diagram are given. The instability strip of the Cepheids in globular clusters is slightly different from those of the Classical Cepheids. Four Cepheids indicated by crosses differ from the other Cepheids in some relations (see e.g. Figure 4).

In Figures 2 and 3 the period-luminosity relations in V and B are given. The above-mentioned four Cepheids are again shown by crosses. The peculiarity of these Cepheids is illustrated by the example of the variables V 60 and V 92 in the globular cluster ω Centauri. Figure 4 gives the light curves of these variables (Martin, 1938). The curve of V 60 is very similar to those of other Cepheids with the same period. The curve of V 92 is very different!

It is possible that these four Cepheids are in phases of evolution different to the majority of Cepheids in globular clusters (Schwarzschild, 1970).

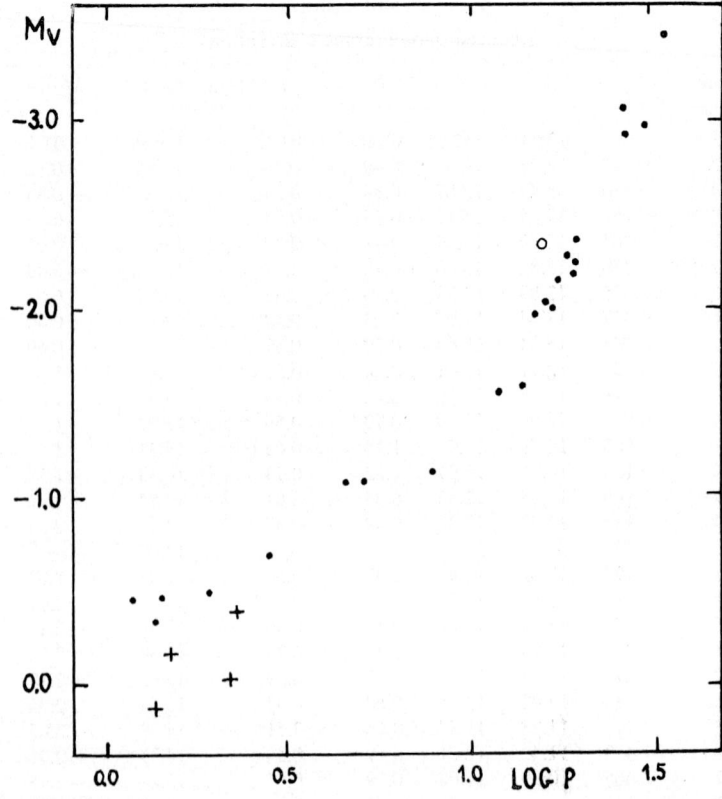

Figs. 2. and 3. The period-luminosity relations for globular cluster Cepheids.

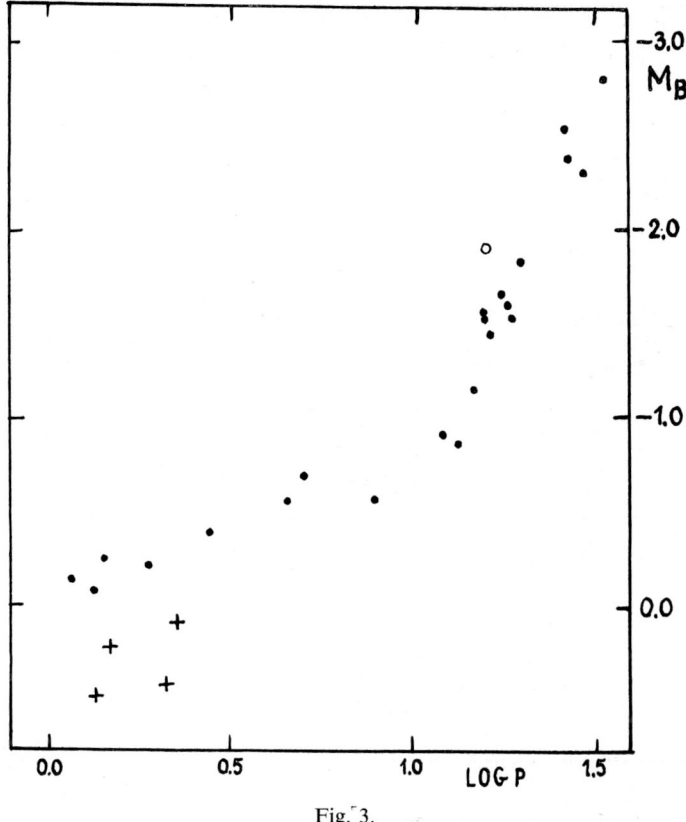

Fig. 3.

The period-luminosity relation of the Cepheids in globular clusters may be represented by the following linear equations:

$$M_V = -0.26 - 1.12 \log P \quad (\log P < 1.14)$$
$$\pm 0.07 \pm 0.08$$
$$M_V = +2.66 - 3.89 \log P \quad (\log P > 1.14)$$
$$\pm 0.10 \pm 0.11$$
(1)

$$M_B = -0.08 - 0.70 \log P \quad (\log P < 1.14)$$
$$\pm 0.06 \pm 0.07$$
$$M_B = -3.51 + 4.11 \log P \quad (\log P > 1.14)$$
$$\pm 0.09 \pm 0.08$$
(2)

The problem of the period-luminosity relation for the Cepheids in globular clusters is complicated. About 85% of Cepheids form a single physical group, but 15% have peculiarities. When using Equations (1) and (2) it is necessary to take into account not only the periods, but also the shape of the light curve and other properties.

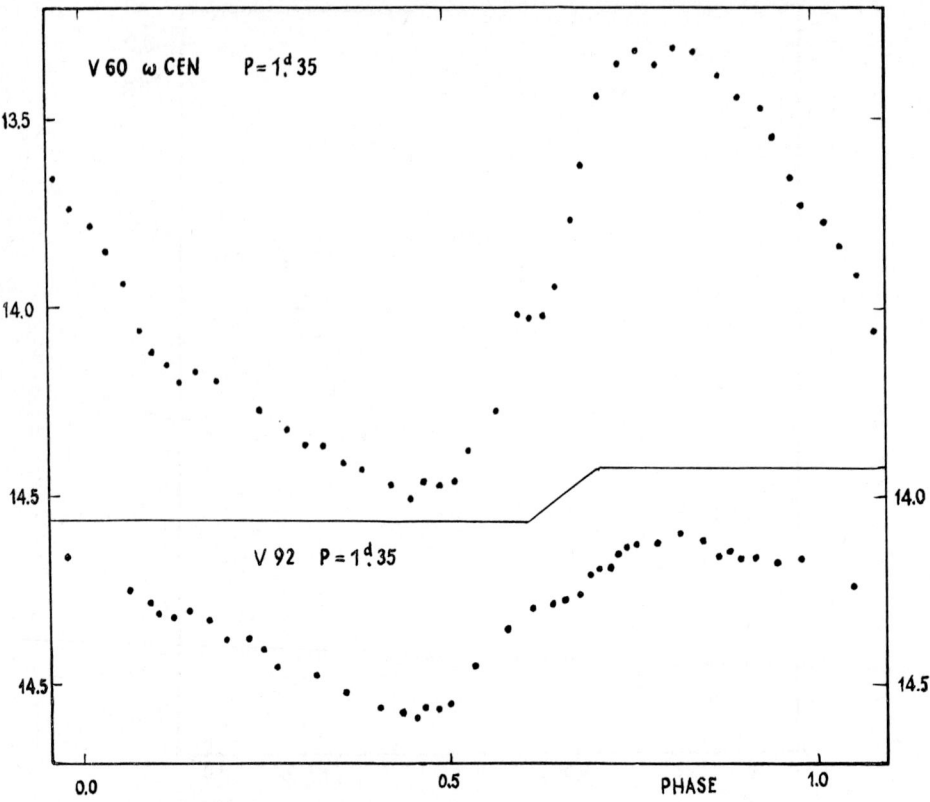

Fig. 4. Differing light-curves of two Cepheids of equal period in ω Cen.

References

Christy, R. F.: 1966, **144**, 108.
Christy, R. F.: 1971, in *Highlights of Astronomy*, Vol. 2, D. Reidel, Dordrecht-Holland, p. 777.
Demers, S.: 1971, *Astron. J.* **76**, 916.
Fernie, J. D.: 1964, *Astron. J.* **69**, 258.
Frolov, M. S.: 1970, in B. V. Kukarkina, (ed.), *Pulsating Stars; Non-stationary Stars and Methods of their Investigation*, Moscow, Izdatel'stvo 'Nauka'. Glavnaya Redaktsiya Fiziko – Matematchesikoy Literatury, p. 124.
Kukarkin, B. V. and Russev, R. M.: 1972, *Astron. Zh.* **49**, 121.
Kwee, K. K.: 1968, *Bull. Astron. Inst. Netherlands* **19**, 374.
Martin, W. C.: 1938, *Ann. Sterrew. Leiden* **17**, No. 2
Pavlovskaya, E. D.: 1953, *Perem. Zvezdy* **9**, 349.
Schwarzschild, M. and Härm, R.: 1970, *Astrophys. J.* **160**, 341.

PART IV

THEORETICAL CONSIDERATIONS OF
POPULATION II VARIABLES

VARIABLE STARS AND EVOLUTION IN GLOBULAR CLUSTERS

PIERRE DEMARQUE

Yale University Observatory, U.S.A.

1. Introduction

Traditionally, cluster variables have been used as distance indicators and have in this sense played an important role in our understanding of stellar evolution. In particular, the determination of the distance moduli of globular clusters and of the absolute magnitude of the main sequence turnoff, thus yielding the ages of the cluster, have relied heavily in the past on observations of RR Lyrae stars.

During the last two decades, it has become possible to follow in some detail the evolution of stars of the halo population from the main sequence to the tip of the giant branch. The last few years have further seen a very rapid advance in our understanding of those phases of evolution which follow the red giant phase, i.e. evolution on the horizontal branch, on the asymptotic branch and directly following the asymptotic branch. We are thus now able to discuss with some reliability the place of the RR Lyrae variables in the evolution of halo stars, and even to discuss the evolutionary status of the long period variables such as the W Virginis stars and the red variables, although perhaps in a somewhat more tentative fashion.

The theory of stellar evolution tells us something about the ages of stars in globular clusters, and about their masses and chemical compositions, these last two parameters to be tested in turn against atmospheric data and the results of detailed calculations of pulsation theory. Theory should eventually explain the existence of the two Oosterhoff (1939) period groups among short period variables. It should also enable us to understand the division recently discussed by Kraft (1972) of the W Virginis stars into two distinct groups, one with periods less than eight days, characterized by the BL Her stars, and the other with longer periods.

Variable stars provide further tests of the theory of stellar evolution. From the calculated rates of evolution it is possible to derive rates of period changes. At the same time, the direction of evolution provides a test of the sense of the period changes, i.e. whether the period is increasing or decreasing. It also gives an explanation for more subtle effects such as that on the relative numbers of Bailey types ab and c variables (Christy, 1966; van Albada and Baker, 1972). This last point will no doubt be discussed at this colloquium by the experts in pulsation theory.

Finally, one should mention the possibility that mass loss is of importance during the RR Lyrae phase, and that it may affect the subsequent evolution of the star. Work on this subject has recently been done by Laskarides (1972) at the University of Victoria.

2. The RR Lyrae Variables

The RR Lyrae variables occupy the center of the horizontal branch. Since the work of Schwarzschild (1940), who showed that all RR Lyrae variables in M3 are located in a well defined color range, and that this color range includes only variables, variability is generally regarded as a property of all horizontal branch stars whose evolutionary tracks take them through this color range. We shall assume here that, in Schwarzschild's words, 'stars which can pulsate, do pulsate', and therefore discuss briefly our present understanding of the horizontal branch.

Although a number of alternative models have been proposed (Hayashi *et al.*, 1962; Larson, 1965; Petersen, 1972), there are good reasons to believe that the structure of horizontal branch stars is the following: a helium burning core surrounded by a hydrogen rich envelope at the bottom of which lies a hydrogen burning shell. Such double energy source models were first constructed by Hoyle and Schwarzschild (1955), but the study of the systematics of the horizontal branch started with the work of Faulkner (1966), since followed by Giannone (1967), by Rood (1970), and by Gross (1972).

A. SYSTEMATICS OF THE ZERO-AGE HORIZONTAL BRANCH (ZAHB)

The systematics of the ZAHB have most recently been studied by Gross (1972), who considered a wide range in heavy element abundances. The four figures have been taken from his work.

Figure 1 shows the effect of adding to a helium configuration of fixed mass a hydrogen rich envelope. If q denotes the ratio of helium core mass to total mass, it is

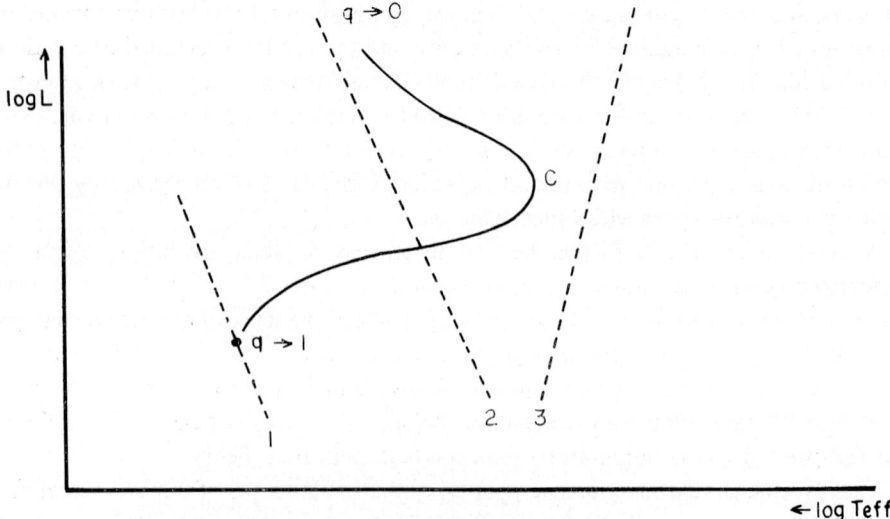

Fig. 1. Locus in the HR diagram of stellar models with constant helium core mass as the ratio q of core mass to total mass varies from 1 to 0.

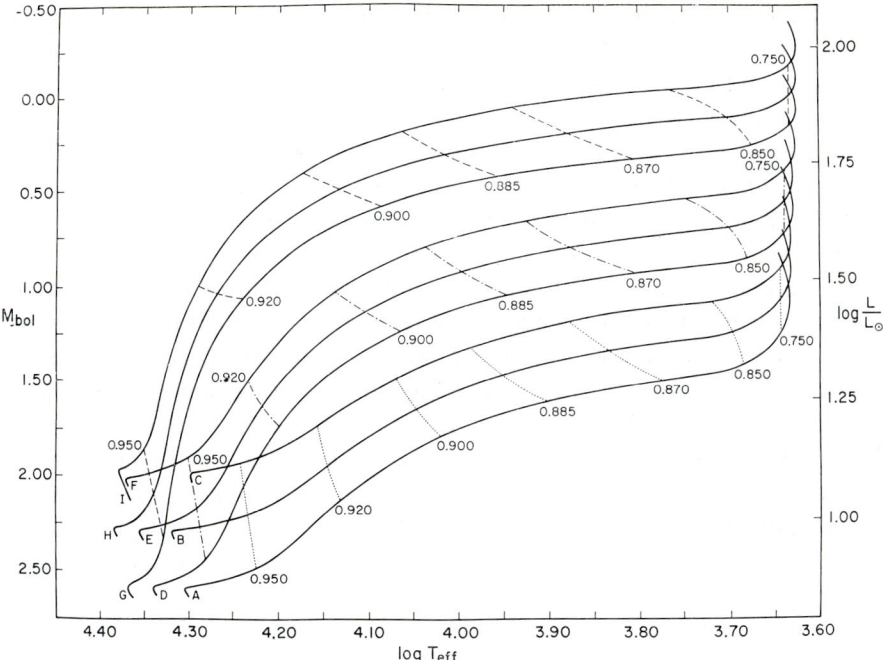

Fig. 2. Position of the ZAHB in the HR diagram as a function of the mass in the helium core and of the initial helium abundance, found in the envelope. If each ZAHB is defined by the parameters $(M_c/M_\odot, Y)$, A corresponds to (0.425, 0.25), B to (0.450, 0.25), C to (0.475, 0.25), D to (0.425, 0.35), E to (0.450, 0.35), F to (0.475, 0.35), G to (0.425, 0.45), H to (0.450, 0.45) and I to (0.475, 0.45). $Z = 0.01$ for all models shown in the figure. Lines of constant q are marked.

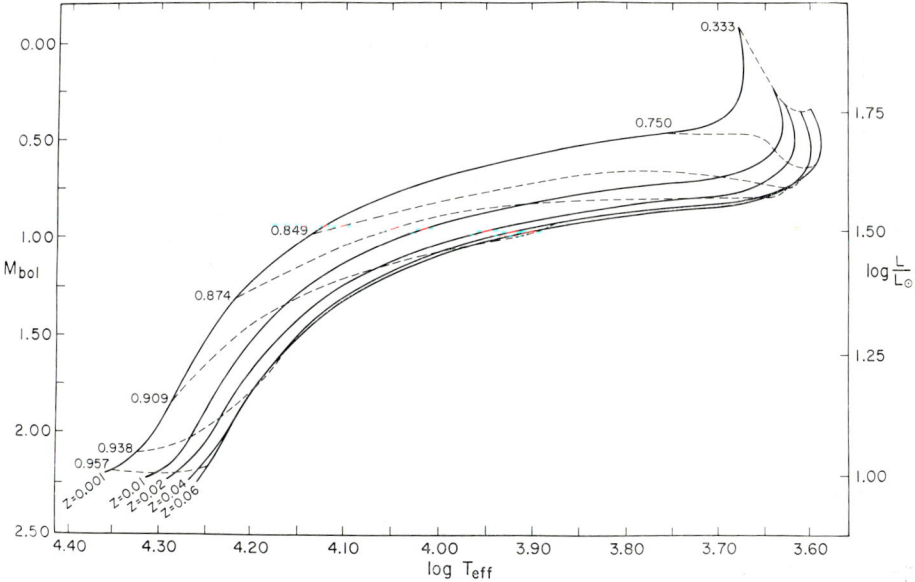

Fig. 3. Position of the ZAHB as a function of Z for $(M_c/M_\odot, Y) = (0.450, 0.25)$. Lines of constant q are marked.

Interpretation (a) was favored by Iben and Rood (1970) in their extensive paper on horizontal branch evolution. They concluded that the observed color spread of horizontal branch stars required a mass range of the order of 0.1–0.2 $M_\odot$ within a given cluster. They further discussed the rates of period changes predicted by theory and found them to be small, at least one order of magnitude less than observed rates quoted in the literature. The subsequent section shows that several important aspects of the Iben-Rood picture must be revised. However, on a number of grounds, it seems that interpretation (a) which they proposed is more likely to be correct than (b).

C. ROLE OF SEMI-CONVECTION

But an important point had escaped most investigators, i.e. the role of semi-convection. Schwarzschild (1970) had already emphasized the occurrence of semi-convection in horizontal branch models. Castellani *et al.* (1971a, b) considered in detail the effects of overshooting from the convective core and the subsequent formation of a semi-convective zone outside the convective core substantially modifying the internal structure of these stars. Evolutionary tracks recently constructed at Yale show that semi-convection plays a significant role in the evolution of horizontal branch stars (Demarque and Mengel, 1972; Sweigart and Demarque, 1972). Ignoring its effects leads to:

(a) too short blue loops in the HR diagram. Recent results indicate that the mass spread needed is less than 0.05 $M_\odot$.

(b) too short horizontal branch lifetimes. The new lifetimes are nearly twice as long as those of Iben and Rood (1970).

The detailed analysis of the effect of overshooting from the convective core into the semi-convective zone further led to the discovery of what appears to be a new 'composition instability', which is described in a subsequent paper (Sweigart and Demarque, 1973). This instability may be of relevance to the study of RR Lyrae stars since it predicts the existence of relatively rapid period changes.

3. Long Period Variables

A. W VIRGINIS STARS

All horizontal branch stars with masses above a certain limiting mass become red giants for a second time, evolving along the asymptotic branch. The internal structure of the star consists of an inert core of carbon and oxygen surrounded by a helium burning shell. A hydrogen burning shell still continues to provide the major part of the luminosity. A major feature of this phase of evolution is the repeated occurrence of a thermal instability in the thin helium burning shell of the star. The history of the star then consists of a series of relaxation cycles including alternating quiescent phases and active phases. The active phases are characterized by one or several brief periods of rapid helium burning. During these shell flashes, peak helium burning rates may exceed 10^7 solar luminosities, while the hydrogen burning shall is practically extinguished (Schwarzschild and Härm, 1965, 1967; Sweigart, 1971; Mengel, 1972). This evolution continues along the asymptotic branch until one of the following two events may occur:

(1) the star reaches such a high luminosity that the ionization instability discussed by Lucy (1967) and by Paczynski and Ziolkowski (1968) takes place in the envelope, probably leading to the loss of the envelope, with or without the help of a shell flash. The star eventually becomes a planetary nebula.

(2) due to its low mass, the star never reaches such a high luminosity. When the envelope mass is of the order of 0.01 $M_\odot$, it starts on its blueward journey across the HR diagram without any appreciable mass loss, at a luminosity level which depends on its total mass and chemical composition.

Both possibilities lead to a single crossing of the W Virginis instability strip from red to blue as the star leaves the asymptotic branch and evolves toward the white dwarf stage.

Further crossings of the W Virginis region may occur during the relaxation cycles (Schwarzschild and Härm, 1970; Mengel, 1972). In most cases, as the star undergoes large changes at the base of the envelope as a result of a helium shell flash, it moves up and down the asymptotic branch. However, in some instances, Schwarzschild and Härm (1970) found that their models left the asymptotic branch and described loops in the HR diagram extending into the W Virginis instability strip. Unfortunately, the original theory predicts period changes which are much more rapid than the available observations would indicate. But slower loops can occur at higher luminosities. A paper by Mengel (1972) at this colloquium gives a discussion of this problem in terms of improved interior models.

It may become possible to test the hypothesis of a spread in total mass on the horizontal branch by studying the W Virginis stars and associated bright non-variables which are found in clusters above the horizontal branch (Zinn *et al.*, 1972). If the mass spread hypothesis is correct, then one would expect a variety of stars to traverse the W Virginis instability strip, principally stars describing slow loops (Mengel, 1972), but also others evolving from the red to the blue on their way to the white dwarf stage. One would then expect a wide spread in luminosities among W Virginis variables and non-variables at the same phase of evolution. If on the other hand all horizontal branch stars have essentially the same mass, then a narrower range of luminosities for the W Virginis variables as well as stars to the blue of the instability strip should be observed within any one cluster.

B. RED VARIABLES

The place of the long-period red variables in the overall evolutionary picture is still unclear. It is possible that at least some of these objects are red giants, i.e. in the pre-horizontal-branch phase of evolution. It seems more likely, however, that they belong to the advanced stages of the asymptotic branch evolution. At any rate, it is tempting to relate these stars to the phase immediately preceding the mass loss process through ionization instability mentioned in the previous section. Theoretical research in this area is presently being conducted at the Princeton Observatory.

Already Smith and Rose (1972) have made hydrodynamic calculations of relaxation oscillations in the envelopes of luminous red giants. They suggest that such oscillations would be driven by high radiation pressure gradients at the base of the envelope, or by instability to non-adiabatic pulsations as recently studied by Keeley (1970).

Acknowledgements

The author's research is supported by Grant GP 21345 from the National Science Foundation. He wishes to express his special thanks to Dr J. G. Mengel and Dr A. V. Sweigart for many discussions of the points covered in this paper.

References

Albada, T. S. van and Baker, N.: 1971, *Astrophys. J.* **169**, 311.
Albada, T. S. van and Baker, N.: 1972, in A. G. Davis Philip (ed.), *The Evolution of Population II Stars*, Dudley Observatory Report No. 4, p. 193.
Cannon, R. D.: 1970, *Monthly Notices Roy. Astron. Soc.* **150**, 111.
Castellani, V., Giannone, P., and Renzini, A.: 1971a, *Astrophys. Space Sci.* **10**, 340.
Castellani, V., Giannone, P., and Renzini, A.: 1971b, *Astrophys. Space Sci.* **10**, 355.
Christy, R. F.: 1966, *Astrophys. J.* **144**, 108.
Cox, A. N. and Stewart, J.: 1970, *Astrophys. J. Suppl.* **19**, 261.
Demarque, P. and Mengel, J. G.: 1971, *Astrophys. J.* **164**, 317.
Demarque, P. and Mengel, J. G.: 1972, *Astrophys. J.* **171**, 583.
Demarque, P. and Mengel, J. G.: 1973, *Astron. Astrophys.*, **22**, 121.
Demarque, P., Mengel, J. G., and Sweigart, A. V.: 1972, *Astrophys. J. Letters* **173**, L27.
Faulkner, J.: 1966, *Astrophys. J.* **144**, 978.
Faulkner, J. and Iben, I., Jr.: 1966, *Astrophys. J.* **144**, 995.
Giannone, P.: 1967, *Z. Astrophys.* **65**, 226.
Gross, P. G.: 1972, Ph.D. dissertation, Yale University.
Hartwick, F. D. A.: 1968, *Astrophys. J.* **154**, 475.
Hartwick, F. D. A., Härm, R., and Schwarzschild, M.: 1968, *Astrophys. J.* **151**, 389.
Hayashi, C., Hoshi, R., and Sugimoto, D.: 1962, *Prog. Theor. Phys. Suppl.*, No. 22, p.1.
Hoyle, F. and Schwarzschild, M.: 1955, *Astrophys. J. Suppl.* **2**, 1.
Iben, I., Jr. and Rood, R. T.: 1970, *Astrophys. J.* **161**, 587.
Keeley, D. A.: 1970, *Astrophys. J.* **161**, 643.
Kippenhahn, R.: 1970, *Astron. Astrophys.* **8**, 50.
Kraft, R. P.: 1972, in A. G. Davis Philip (ed.), *The Evolution of Population II Stars*, Dudley Observatory Report No. 4, p. 69.
Larson, R. B.: 1965, *Publ. Astron. Soc. Pacific* **77**, 452.
Laskarides, P. G.: 1972, Ph.D. dissertation, University of Victoria.
Lucy, L. B.: 1967, *Astron. J.* **72**, 813.
Mengel, J. G.: 1972, Ph.D. dissertation, Yale University (see also this volume, p. 214).
Oosterhoff, P. Th.: 1939, *Observatory* **62**, 104.
Paczynski, B. and Ziółkowski, J.: 1968, *Proc. IAU Symposium No. 34 on Planetary Nebulae*.
Petersen, J. O.: 1972, *Astron. Astrophys.* **19**, 197.
Rood, R. T.: 1970, *Astrophys. J.* **161**, 145.
Schwarzschild, M.: 1940, *Harvard Circ. No. 437*.
Schwarzschild, M.: 1970, *Quart. J. Roy. Astron. Soc.* **11**, 12.
Schwarzschild, M. and Härm, R.: 1962, *Astrophys. J.* **136**, 158.
Schwarzschild, M. anb Härm, R.: 1965, *Astrophys. J.* **142**, 855.
Schwarzschild, M. anb Härm, R.: 1967, *Astrophys. J.* **150**, 961.
Schwarzschild, M. and Härm, R.: 1970, *Astrophys. J.* **160**, 341.
Smith, R. L. and Rose, W. K.: 1972, *Astrophys. J.* **176**, 395.

Sweigart, A. V.: 1971, *Astrophys. J.* **168**, 79.
Sweigart, A. V. and Demarque, P.: 1972, *Astron. Astrophys.*, **20**, 445.
Sweigart, A. V. and Demarque, P.: 1973, this volume, p. 221.
Zinn, R. J., Newell, E. B., and Gibson, J.: 1972 *Astron. Astrophys.* **18**, 390.

DISCUSSION

Belserene: About the time scale for evolution through the instability strip, you said that the theorists need longer times than the ones we find. Are you referring to blueward evolution? Can you accept 2 to 10 million years as reasonable in a cluster with a very blue horizontal branch, where the only RR Lyr stars are the ones we catch on their way to the asymptotic branch.

Demarque: Yes, it is conceivable, depending on how blue the star was originally.

Walborn: If the horizontal branch is interpreted as a mass sequence, are the RR Lyrae stars then stars of a certain mass rather than stars at a certain evolutionary stage?

Demarque: Neither. Stars of different evolutionary stages will be found in the instability strip at any given time.

van den Bergh: Recent work by Peimbert seems to indicate that the oxygen-to-iron ratio might differ from cluster to cluster. Could you tell us how this might affect the structure of the horizontal branch? Could this be the 'second parameter' that seems to be required to account for the differences in the observed population gradients along the horizontal branch of clusters with the same [Fe/H]?

Demarque: Yes, the colours of horizontal branch stars depend most critically on the strength of the hydrogen burning shell, which depends directly on the abundance of CNO elements. Opacity effects are less important.

Lloyd Evans: (1) If the flash occurs at a luminosity below the observed tip of the giant branch shouldn't we see a discontinuity in luminosity function at the position of the flash? (2) The suggestion that some red variables, hopefully losing mass, are at the flash stage of evolution, does not seem to be much help in explaining the low mass of blue HB stars since the most pronounced red variables (Miras especially) are found in just those (metal rich) clusters with only the red stub horizontal branch.

Demarque: (1) The rate of evolution on the asymptotic branch is roughly one third that on the giant branch. In view of the small numbers of stars at this luminosity level, it would be difficult to observe this effect. (2) Little can be said about the masses of the stars on the red stub if the metal abundance is high. It is possible that such stars may have lost mass on the giant branch.

Schwarzschild: Could Dr Baker give us his view regarding the plausible mass or mass-range for the horizontal branch? I would feel that spectroscopic determinations do not yet reach the relatively high accuracy here required. Next, I still feel rather doubtful that finite-amplitude pulsation theory – as much as I admire its successes – can yet decisively contribute to this question. There remains then the general position of the horizontal branch in the HR diagram; can it give fairly definite masses for the horizontal branch?

Demarque: Yes, interior models give masses in the range 0.55–0.65 $\mathfrak{M}_\odot$. Recent computations by Sweigart and myself indicate that on the basis of the observed spread in colour on horizontal branches, there is a mass spread of about 0.05 $\mathfrak{M}_\odot$ in a given cluster.

Cox: In connection with this question of whether the horizontal branch can be explained without any mass loss, could you again review what would happen to a star undergoing a helium flash in the case of no (or very little) mass loss? Is this an acceptable situation?

Demarque: If we adopt a mass of 0.8 $\mathfrak{M}_\odot$ for the main sequence turnoff, and there is no mass loss on the giant branch, present models predict no horizontal branch, but rather a red clump similar to that found in old clusters of the disk population.

Feast: If I understood you correctly, you said that it may be necessary to consider that the various red variables, W Virginis, and RV Tauri stars in clusters can only be understood if there is a range of masses. Might one then expect that a cluster which contains a good selection of these variables would also show a larger than normal scatter of non-variables around the red giant branch? This appears to be the case, for instance, in ω Cen.

Demarque: I think personally that recent evolutionary calculations, in particular the work of Mengel, favor the hypothesis of a mass spread. But the situation is far from settled. I agree with your remark on ω Cen.

ON THE TWO OOSTERHOFF GROUPS OF GLOBULAR CLUSTERS

T. S. VAN ALBADA

Kapteyn Laboratory, Groningen University, Groningen, The Netherlands

and

NORMAN BAKER

Astronomy Dept., Columbia University, New York, N.Y., U.S.A.

Abstract. The observational evidence leading to the classification, following Oosterhoff, of globular clusters containing RR Lyrae stars into two distinct groups, is summarized and discussed in the light of results of stellar evolution theory and pulsation theory. The dichotomy is caused, at least in part, by a dichotomy in the 'transition period' between the type-*ab* and type-*c* stars which reflects a difference in effective temperature at the transition point. When this difference is accounted for, there remains a smaller average difference between the groups, though no longer a clear dichotomy, that is probably a mass and luminosity effect. If this remaining difference is interpreted as a luminosity effect the average difference in luminosity between the two Oosterhoff groups is at most 0.1 mag. It is suggested that Christy's theoretical relationship between transition period and luminosity cannot be valid, at least not for clusters of different Oosterhoff groups. It is conjectured that the transition-temperature dichotomy may be a reflection of different predominant directions of evolution along the horizontal branch, accompanied by a hysteresis effect in the pulsations.

DISCUSSION

Dickens: A significant objection to the hypothesis of a mass difference to explain the dichotomy is the fact that the stellar models of lower mass are bluer (corresponding to clusters with blue horizontal branches) than those of higher mass (red horizontal branches), in the opposite sense to that required to explain the dichotomy in this way. Although changing the metal content can also strongly affect the horizontal branch models, the fact that it correlates rather poorly with horizontal branch type implies that a variation in the mass is likely to remain as an important parameter in interpreting the colour-magnitude diagrams of clusters. This appears to strengthen the above objection to a mass difference between clusters of different Oosterhoff groups.

ON THE INTERPRETATION OF RR LYRAE PROPERTIES IN GLOBULAR CLUSTERS AND IN OTHER POPULATION II SYSTEMS

V. CASTELLANI

Istituto di Fisica, Università di Roma; Laboratorio di Astrofisica Spaziale, Frascati, Italy

P. GIANNONE

Osservatorio Astronomico, Roma, Italy

and

A. RENZINI

Osservatorio Astronomico, Bologna, Italy

Abstract. The differences in observational parameters of the RR Lyrae variables and horizontal branch stars of globular clusters and other population II systems are considered. A discontinuous behaviour of some parameters is outlined. The Oosterhoff dichotomy and the HB morphology are discussed with regard to a conjecture of mass loss in the pre-HB phase.

1. Introduction

The main problem concerning Population II systems is to determine the age and the helium abundance of each of them. Knowledge of these parameters for a large number of Population II systems has important consequences for cosmology (e.g. the interpretation of the Hubble time, the problem of the helium genesis, the formation of galaxies, and in particular of our own Galaxy and its early chemical evolution).

Unfortunately, the absolute determinations of age and helium abundance from the theory of stellar evolution involve serious difficulties. At present, the uncertainty in age can be estimated to be about 40% when the uncertainty in the theoretical models is also taken into account (Renzini, 1971). The situation regarding the determination of the helium content is no better. Available estimates of $Y = 0.29 \pm 0.03$ for globular clusters (e.g. Iben, 1971) are based on HB evolutionary models with improper fitting conditions at the boundary of the convective cores.

The existence of a semiconvective region surrounding the fully convective core (Schwarzschild, 1970; Castellani *et al.*, 1971; Demarque and Mengel, 1972) nearly doubles the core helium-burning lifetime. This fact leads to a helium abundance of about 0.15 (using the same procedure as Iben, see also Demarque *et al.*, 1972). Moreover, since recent computations of red giant evolutionary models (Rood, 1972) lead to RG lifetimes shorter than previous values by about 15 or 20%, the estimated helium abundance is further reduced by about 0.03.

Therefore, in our opinion, a standard error of ± 0.03 in the helium determination is a rather optimistic estimate. Owing to uncertainties in opacity, mass of the helium core at the flash, treatment of semiconvection and stellar counts, we believe that at present the uncertainty in the determination of the helium content of Population II stars might even be ±0.15. This latter estimate still allows the possibility of a high helium content ($Y \simeq 0.30$).

The failure of accurate *absolute* determinations of age and helium abundance of globular cluster stars emphasizes the importance of *relative* determinations of these parameters. This procedure should enable us to answer the question "how large, if any, are age and helium differences among the various Population II systems?".

Within this framework it becomes particularly important to consider the differences in the observational parameters of the Population II systems. In fact there is a popular tendency to assign to all the galactic globular clusters the same age and the same helium content, in spite of the growing evidence that something else besides the metal content varies from cluster to cluster (van den Bergh, 1967; Sandage and Wildey, 1967; Hartwick, 1968; Castellani *et al.*, 1970; King, 1971; Dickens, 1972). Therefore, we cannot rule out the possibility that the Population II systems have different ages and/or helium contents, at least until both observation and theory prove that the relative abundances of metals are the so called 'second parameter' (e.g. Z_{CNO}/Z).

This communication is an account of a rather larger paper in preparation (hereinafter referred to as CGR 72) dealing with the interpretation of observed differences among Population II systems. We now wish to discuss the problem of the well-known Oosterhoff dichotomy in relation to the morphology of the horizontal branches of globular clusters and related systems. Indeed, a 'good theory' should succeed in explaining RR Lyrae properties and HB morphologies at the same time.

Observational evidence concerning RR Lyrae variables can be used in the framework of pulsation theory. However, according to the current literature, many possible explanations of the observations (e.g. the Oosterhoff effect) exist. On the other hand, observed morphologies of horizontal branches can be related to evolution theory. However, even in this case some ambiguities remain.

When the requirements corresponding to the two separate attempts are fitted together, a number of possible explanations have to be ruled out because they don't fit either the pulsation theory or the HB evolution theory. Some remaining ambiguities can be removed when the results of the evolution prior to the HB phase and the present knowledge and belief about the process of formation of globular clusters and related systems are considered.

It may be that the procedure just described is not yet able to eliminate all the ambiguities and to yield firm conclusions because of observational and theoretical uncertainties. However such an approach might assist in clarifying the problem and suggest which observations and computations will be the more important.

Furthermore, we feel that a fully quantitative theory is not quite meaningful at present and that a qualitative, or a semi-quantitative, approach is preferable. For instance, one has that, according to pulsation theory, the difference in the mean period of the *ab*-type variables between the two Oosterhoff groups may be ascribed to a luminosity difference $\Delta \log L \simeq 0.1$, or to a mass difference $\Delta \log M \simeq -0.1$ or to a difference in the mean effective temperature $\Delta \log T_{eff} \simeq -0.02$ (corresponding to $\Delta(B-V) \simeq 0.07$), or, finally, to a combination of these three quantities. It is also worthwhile to note that the foregoing numbers are very close (particularly in the first two cases) to the uncertainties which affect any direct determination of absolute luminosities, masses and colours of HB stars.

In the following we shall emphasize in the whole pattern of the observations: (i) continuity or discontinuity in the observed parameters and, (ii) morphology of the horizontal branches, i.e. we shall be concerned with the topological properties of the observations. From a theoretical point of view, we shall try to find a theoretical scheme topologically analogous to the observational picture.

2. Observational Evcidence for the RR Lyrae Variables in Galactic Globular Clusters

In this section the most relevant charecteristics of the so called Oosterhoff effect are summarized. A longer discussion will appear in CGR 72.

In passing from one Oosterhoff group to the other, we have tried to determine whether each observational parameter changes continuously or discontinuously. In the latter case, the amount of the discontinuity was obtained on the basis of the distribution of the available observational data (rather than as a difference between the mean values corresponding to the two Oosterhoff groups). Bearing this procedure in mind we have the following indications:

(i) The mean period of the ab-type variables suffers a discontinuity $\Delta \langle \log P_{ab} \rangle \simeq 0.07$ (see Figures 1 and 2).

(ii) The mean 'fundamental period – the mean period of all the RR Lyrae variables in a cluster when the periods of the c-type variables are transformed to the fundamental

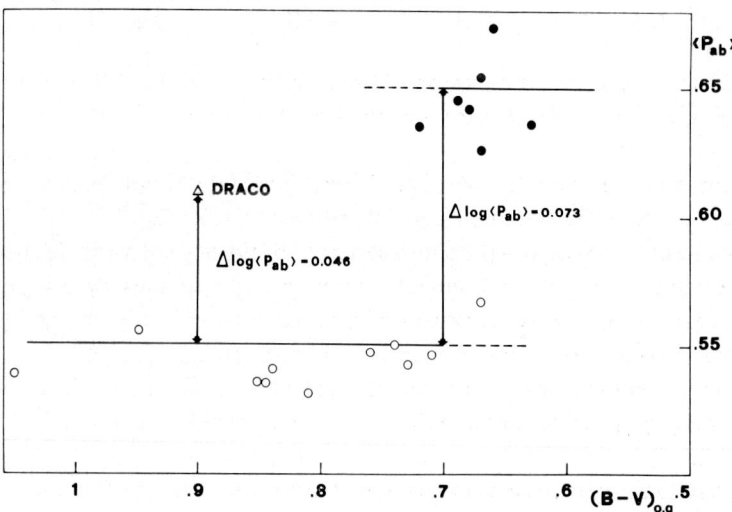

Fig. 1. The mean period of the ab-type variables of Oosterhoff type I and II clusters and of the Draco system is plotted against $(B-V)_{0,g}$. Differences in $\Delta \log \langle P_{ab} \rangle$ between the two Oosterhoff groups and with respect to Draco are also indicated.

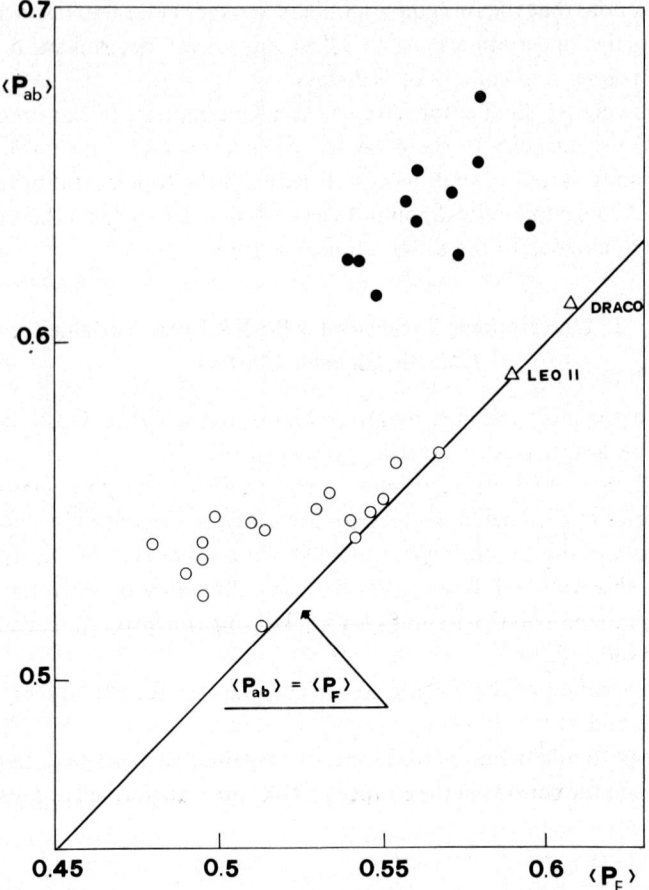

Fig. 2. The mean period of the *ab*-type variables of Oosterhoff type I and II clusters and of the dwarf galaxies Draco and Leo II is plotted against the fundamental period (see text).

period – turns out to be continuous, i.e. $\Delta \langle \log P_F \rangle \simeq 0$ (see Figure 2). A discussion of the incidence of some selection effects will appear in CGR 72.

(iii) The transition period – the shortest period of the *ab*-type variables in a cluster (in some cases very poorly determined) – may be discontinuous by $\Delta \log P_{tr} \lesssim 0.07$.

(iv) The median period – the period of a variable lying just in the middle of the instability strip – may be discontinuous by $\Delta \log \tilde{P} \lesssim 0.07$.

(v) The relative number of the *c*-type variables to the *ab*-type variables is discontinuous by $\Delta (n_c/n_{ab}) \simeq 0.5$, if one excludes three clusters classified as CII (Castellani *et al.*, 1970).

(vi) The instability strip seems to be populated uniformly in type II clusters, whereas there is a strong departure from uniformity in type I clusters (see CGR 72).

(vii) No significant trend in $\langle \log P_{ab} \rangle$ is seen within each Oosterhoff group (see for instance Figure 2 and Figure 5 in Castellani *et al.*, (1970)). This means that very likely

$\partial \langle \log P_{ab} \rangle / \partial \log Z \simeq 0$. A similar statement must hold for the dependence of $\langle \log P_{ab} \rangle$ upon a possible 'second parameter'.

On the basis of the foregoing considerations, it seems possible to conclude that a real discontinuity, rather than a difference between two widely separate classes of objects, exists between the two Oosterhoff groups.

3. The HB Morphology and Evolution

Dickens (1972) classifies the morphological appearance of the horizontal branches of globular clusters into seven classes according to the colour distribution of the HB stars. Dickens' class 1 corresponds to a completely blue HB, without RR Lyrae variables or red HB stars (prototype M13). Dickens' class 7 corresponds to a stubby red HB as in 47 Tuc. Oosterhoff type II clusters are confined essentially to Dickens' classes 2 and 3, whereas Oosterhoff type I clusters are contained in Dickens' classes 3 to 6.

It turns out that all the Oosterhoff type II clusters have horizontal branches well populated on their blue sides and lacking or only poorly populated on their red sides. On the contrary, the Oosterhoff type I clusters always have at least some red HB stars and exhibit an extremely variable population on the HB blue side (compare for instance M5 and NGC 362).

At first sight it is rather surprising that the clusters in Dickens' class 1 are, on the average, more metal-rich than the clusters in class 2 and more metal-poor than the clusters in class 3 and in the further ones. The meaning of the Dickens classification should be the following. In passing from the Oosterhoff type II clusters to clusters like 47 Tuc, at increasing metal abundances, the well populated part of the HB moves first to bluer colours, leaving a depopulated instability strip. Then, by further increasing the metal abundance, the blue end of the HB becomes even bluer or remains fixed; the instability strip and the red side of the HB begin to be populated. The increasing population of the red side of the HB is not accomplished by a shift of the well populated part of the HB towards the red, but by a progressive depopulation of the blue side. Eventually, no blue HB star remains and one has a stubby red HB.

It is well known that considerable difficulties are encountered when one tries to fit the observed horizontal branches with the available evolutionary tracks. This holds even if semiconvection effects are taken into account, and in particular if one assumes a constant stellar mass evolving from the main sequence to the HB.

The most relevant of these difficulties are the following:

(i) The bluest HB stars are not fitted by stellar models unless ages are exceedingly large.

(ii) The effective temperature range of the slowly travelled parts of the evolutionary tracks is usually too small compared with the observed range of HB colours.

(iii) Single evolutionary lines don't usually fit the observed gradients in the effective temperature distribution of the stars along the HB.

(iv) If only the metal content is allowed to change from cluster to cluster, it is not possible to reproduce all the observed features of the horizontal branches.

One way out of these difficulties is to reject the theoretical evolutionary tracks computed up to now, and to hope that the fitting will improve with better input physics. The second way is to relax some of the foregoing assumptions, eventually allowing for some spread in the stellar parameters and for some mass loss prior to the HB phase.

Although a certain spread in chemical composition or in helium-core mass among the HB stars in a cluster may perhaps produce analogous effects, the simplest assumption seems to be that of allowing for a loss of mass in the pre-HB phase. In this case, the bluest HB stars can be fitted (Castellani and Renzini, 1968). If some spread in the mass loss for the HB stars of a cluster is also assumed, it is not difficult to account for the observed range of effective temperatures (Iben and Rood, 1970; Castellani et al., 1970; Iben, 1971, 1972).

In this framework, point (iii) seems to require different HB mass functions in different clusters, and point (iv) indicates that the mean amount of the mass loss in a cluster must be very sensitive to the metal abundance and to the 'second parameter'.

This set of assumptions will be referred to in the following as 'the mass loss conjecture'. In the subsequent section, the consequences of the mass loss conjecture will be compared with the requirements concerning the behaviour of the HB parameters induced by considering the Oosterhoff effect.

4. The Oosterhoff Effect and the HB Morphology

According to pulsation theory, the discontinuity in $\langle P_{ab} \rangle$ between the two Oosterhoff groups implies a discontinuity either in the mean luminosity of the ab-type RR Lyrae variables, or in their mean mass, or in their mean effective temperature, or finally a more sophisticated combination of $\Delta \langle \log L \rangle$, $\Delta \langle \log M \rangle$ and $\Delta \langle \log T_{\text{eff}} \rangle$.

If one follows the mass loss conjecture (or more generally assumes a certain range in the HB stellar parameters) it turns out that a discontinuity between the two Oosterhoff groups must occur in at least one of the mean values of the four parameters which control the HB evolution (Z, Y, M_{core} and M). In the following we shall discuss in some detail the four simplest possibilities, i.e. the discontinuity in $\langle P_{ab} \rangle$ is due to a discontinuity in Z, or in Y, or in M_{core}, or finally in M. Furthermore, a discontinuity in some of these four parameters may be due either to an initial difference in age or chemical composition between the two Oosterhoff groups, or to the existence of some threshold mechanism operating during the pre-HB evolution.

A discontinuity in Z can be ruled out on two grounds. The observational evidence indicates that, actually, metallicity varies continuously in passing from one Oosterhoff group to the other; the mean period of the ab-type RR Lyrae variables appears to be independent of metallicity within an Oosterhoff group.

A discontinuity in Y might be due either to an initial difference in the helium content, or to some threshold mechanism for a partial mixing of the superficial stellar layers. Very little can be said about the first possibility; up to date pre-HB evolutionary computations seem to exclude the second eventuality.

A discontinuity in M_{core} has been considered for some time as a possibility related to the nitrogen flash. However, recent computations of the cross-section for the nitrogen α-capture (Couch, *et al.*, 1972) exclude the possibility of the nitrogen flash actually occurring. Therefore it seems very unlikely that a discontinuity in M_{core} is the cause of the Oosterhoff dichotomy.

The possibility of a discontinuity in the stellar mass is in accordance with the mass loss conjecture. A threshold mechanism for the mass loss should only act for stars with stellar parameters in a certain range. Then a 'critical boundary' should mark the transition corresponding to the discontinuity in the amount of mass loss.

It is quite natural to locate this hypothetical threshold mechanism at the helium flash. The discontinuous mass loss occurring at the flash might be superposed on a steady loss of matter taking place during the red giant evolution. Very likely the stars in a cluster are not strictly identical; core rotation, magnetic fields, chemical differences could lead to a certain range of core and stellar masses at the helium flash.

As a result, stars at the flash would be characterized by a finite dispersion in a diagram with pertinent stellar parameters as coordinate axes. A variety of different situations can then arise according to the relative position of the group of representative star points and of the critical boundary. In Figure 3 are plotted the morphologies

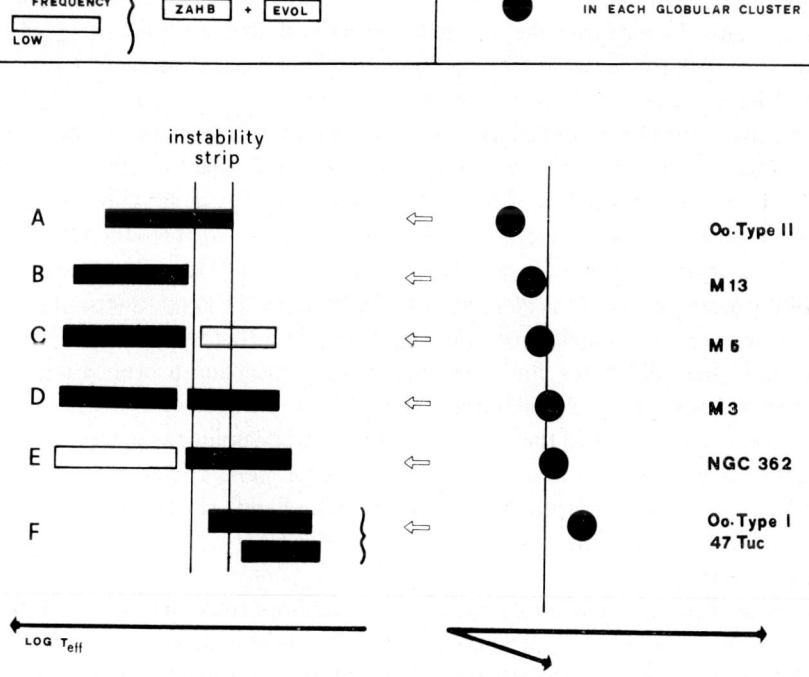

Fig. 3. Schematic morphologies in the HR diagram of horizontal branches of globular clusters (left-hand panel). They are related to the relative locations in a pertinent diagram of representative star points and of the 'critical boundary' for the occurrence of mass loss (right-hand panel).

of the horizontal branches as they result from various cases, schematically represented by the mutual locations of stars of a cluster (circles) and of the critical boundary (vertical straight line).

In case A, all the stars in the cluster suffer a sudden mass loss; correspondingly the slowly travelled part of the HB is blue as in the clusters in Dickens' class 2. For an increasing metal abundance one would expect an increasing mass loss (the surface gravity is smaller along the red giant branch). The HB evolutionary tracks shift to bluer colours and move out of the instability strip (case B, as in Dickens' class 1). When the metal abundance increases further, a few stars begin to cross the critical boundary. Then the threshold mechanism ceases to act for them; they do not lose mass, and they start to populate the red side of the HB and the instability strip. On the other hand, most of the stars continue to lose mass and to populate the blue side of the HB (case C). This situation corresponds to Dickens' class 3, whose prototype is M5.

As the sample of representative star points is shifted across the critical boundary because of increasing metal abundance, more and more stars stop losing mass. Correspondingly, the HB appears to be populated more and more uniformly on both sides of the instability strip (case D, resembling Dickens' class 4 with prototype M3). Continuing the trend already outlined, when only a few stars still lose matter, there is a progressive depopulation of the blue side of the HB in favour of the red side and the instability strip (case E, corresponding to Dickens' classes 5 and 6). Finally, no stars undergo mass loss by means of the threshold mechanism, and one has only a red HB with few or no RR Lyrae variables (mostly ab-type). This situation (case F) corresponds to Dickens' class 7, whose prototype is 47 Tuc.

With regard to the Oosterhoff effect, case A corresponds to Oosterhoff type II clusters, cases C, D, E, and in part F, correspond to Oosterhoff type I clusters. The foregoing picture then reproduces quite closely the observational evidence that the Oosterhoff type II clusters always have blue horizontal branches, whereas the Oosterhoff type I clusters exhibit a large variety of HB morphologies. In this scheme, the discontinuity in the mean period of the ab-type variables between the two Oosterhoff groups is due primarily to a mass difference ($\Delta \log M \simeq -0.1$, according to pulsation theory). However, this mass difference could be smaller if the mechanism proposed by van Albada and Baker (1972) is actually operating.

Up to now we have assumed that globular clusters form a unique sequence. Actually this is not the case. Clusters believed to have the same metal content have in many cases quite different HB morphologies. On the other hand, clusters believed to be rather metal-different exhibit similar HB morphologies. This fact is related to the 'second parameter' puzzle.

The hypothetical threshold mechanism for the mass loss could act as an 'amplifier' of small differences in the parameters other than the metal content. Indeed it is very likely that the helium flash phenomenon depends primarily on the core mass. An equal difference in the core mass is produced either by a factor ten in the metal content, or by a difference of about 0.05 in the helium content (Rood, 1972). Therefore, if the helium abundance is the second parameter, a difference $\Delta Y \leq 0.05$ among the globular clusters is sufficient to lead to rather large effects as far as the HB morphology is concerned.

As is well known, the dwarf galaxies Draco and Leo II show a mean period of the *ab*-type variables intermediate between those of the two Oosterhoff groups. In Figure 2 Draco and Leo II lie in a region which is lacking in galactic globular clusters. This means that an intrinsic parameter (constant in the clusters) has different values in the two galaxies.

An age difference of a few billion years could explain the strange behaviour of the RR Lyrae variables in these two latter systems. Indeed, the HB mass decreases secularly at a rate of about 2 percent per billion years because of evolutionary effects. As a consequence of that, the zero-age HB location shifts bluewards at a rate of about $\Delta \log T_{\text{eff}} \simeq 0.02$ per billion years. Correspondingly, the part of the instability strip where long period variables are located is depopulated and the mean period decreases. In order to have the observed value of $\langle P_{ab} \rangle$, it would be sufficient if Draco and Leo II were younger than the galactic globular clusters by about 1 to 3 billion years.

References

Albada, T. S. van and Baker, N. H.: 1972, *The Evolution of Population II Stars*, Dudley Observatory Report No. 4, p. 193.
Bergh, S. van den: 1967, *Astron. J.* **72**, 70.
Castellani, V., Giannone, P., and Renzini, A.: 1970, *Astrophys. Space Sci.* **9**, 418.
Castellani, V., Giannone, P., and Renzini, A.: 1971, *Astrophys. Space Sci.* **10**, 355.
Castellani, V. and Renzini, A.: 1968, *Astrophys. Space Sci.* **2**, 310.
Couch, R. G., Spinka, H., Tombrello, T. A., and Weaver, T. A.: 1972, *Astrophys. J.* **172**, 395.
Demarque, P. and Mengel, J. G.: 1972, *Astrophys. J.* **171**, 583.
Demarque, P., Sweigart, A. V., and Gross, P. G.: 1972, preprint.
Dickens, R. J.: 1972, *Monthly Notices Roy. Astron. Soc.* **157**, 281.
Hartwick, F. D. A.: 1968, *Astrophys. J.* **154**, 475.
Iben, I.: 1971, *Publ. Astron. Soc. Pacific* **83**, 697.
Iben, I.: 1972, *The Evolution of Population II Stars*, Dudley Observatory Report No. 4, p. 1.
Iben, I. and Rood, R. T.: 1970, *Astrophys. J.* **159**, 605.
King, I. R.: 1971, *Publ. Astron. Soc. Pacific* **83**, 377.
Renzini, A.: 1971, *Mem. Soc. Astron. Ital.* **42**, 349.
Rood, R. T.: 1972, preprint.
Sandage, A. and Wildey, R.: 1967, *Astrophys. J.* **150**, 469.
Schwarzschild, M.: 1970, *Quart. J. Roy. Astron. Soc.* **11**, 13.

DISCUSSION

van den Bergh: Available evidence suggests that the rate of star formation in a Galaxy $\sim \varrho^n$, in which ϱ is the initial gas density. This explains why *E* galaxies, in which ϱ is high contain mostly old stars. Irregular galaxies, which have a low mean density, have not yet used up all their interstellar gas and consequently contain many young stars. The mean density of dwarf spheroidal galaxies is even lower than that in irregulars. This seems to conflict with the observation that dwarf spheroidals contain only old stars. This suggests that dwarf spheroidals were not formed as individuals but were created together with giant galaxies. Observations of the distribution of dwarf spheroidals in the M81 group, the South Pole Group and the Local Group give strong support to the idea that dwarf spheroidals are associated with giants. Presumably this implies that the dwarf spheroidals formed (like globular clusters) during the collapse of the giant galaxy with which they are associated. On such a picture the age difference between the Draco System and the Galaxy cannot exceed the collapse time-scale of the Galaxy, which is only a few times 10^8 yr. It seems doubtful if such a small age difference can account for the differences that are observed between the dwarf spheroidal satellites of the Galaxy and galactic globular clusters.

Renzini: What I wanted to show is that the distribution of stars along the horizontal branch may be rather sensitive to age. An age difference of 1 to 3 billion years would produce the observed difference in $\langle P_{ab} \rangle$ between the Leo II and Draco systems on the one hand and the galactic globular clusters of Oosterhoff type I on the other hand. Nevertheless, this age difference is only a *sufficient*, not a *necessary*, condition. It may be that a difference in some other parameter is responsible for the strange values of $\langle P_{ab} \rangle$ in the two dwarf galaxies. However, this hypothetical parameter cannot be the same 'second parameter' invoked for the galactic globular clusters, otherwise no dichotomy would be observed in $\langle P_{ab} \rangle$ among the globular clusters.

Demarque: Are you appealing to small differences in helium content from cluster to cluster, or from star to star within the cluster?

Renzini: What I appeal to within one cluster is a certain range in the stellar parameters, more likely due to different core rotations. A range in helium among the stars in one cluster cannot be excluded, but I don't strictly require that. A small range in helium from cluster to cluster would instead be responsible for the different appearance in horizontal branches in clusters with about the same metal content.

POSSIBLE EVOLUTIONARY INTERPRETATION OF THE DEPENDENCES ON THE DIAGRAMS FOR RR LYRAE VARIABLES

V. P. GORANSKIJ
Moscow University, Moscow, U.S.S.R.

The separation of RR_{ab} variables into two sequences on the amplitude-period diagram was found by Belserene (1954) for globular cluster variables, and by Detre (1965) for field variables. It is significant that the sequence of short period variables is steeper on this diagram than that for stars of longer period. Szeidl (1965) pointed out that other characteristics of the light curves and photometric quantities such as the asymmetry of light curve ε, mean color index $B-V$, median magnitude m_{med}, maximum magnitude m_{max} are also separated into distinct sequences. Figures 1, 2 and 3 show these sequences on the colour-magnitude, colour-period, period-amplitude (or period-maximum magnitude, minimum magnitude) diagrams for RR_{ab} variables in the globular clusters M3, ω Cen and in the dwarf galaxy in Draco. On these figures RR_{ab} variables of the long period sequence are marked by crosses. The colour-magnitude and colour-period diagrams for M3 were obtained by Roberts and Sandage (1955); the period-maximum magnitude, minimum magnitude diagram obtained by Szeidl (1965) is constructed for stars measured by Roberts and Sandage. To get the most accurate V-magnitudes and colour indices for variables in ω Cen the data of Dickens and Saunders (1965) and Geyer and Szeidl (1970) have been averaged. The diagrams for RR Lyr variables in the Draco system were obtained by Baade and Swope (1961). The variables which show the Blazhko effect were excluded.

One can see on the diagrams for RR Lyrae variables that the long period sequence stars which depart from the basic sequences have a systematically higher luminosity than ordinary stars. The RR_c variables of increased luminosity (crosses) also have systematically greater periods.

This complex of phenomena on the diagrams for RR Lyrae variables can be treated by means of calculations of the evolution of models with core helium burning (Iben and Rood, 1970) and pulsation theory (Christy, 1966).

Iben and Rood calculated the evolution of low-mass models (0.6–0.7 $\mathfrak{M}_\odot$) with helium burning in a core and hydrogen burning in a shell. An example of the evolutionary tracks of such models is illustrated in Figure 4. The parameters of the models are shown on the figure. ZAHB (zero age horizontal branch) is a mass sequence of the initial models with a core mass of 0.475 $\mathfrak{M}_\odot$. The separation between two adjacent marks along each track corresponds to a time interval of 5×10^6 yr. According to the calculations the star evolves to the blue on the HR diagram during the most prolonged stage of its life on the horizontal branch. The surface temperature of the star increases during this stage and the radius decreases. The luminosity of the star remains approx-

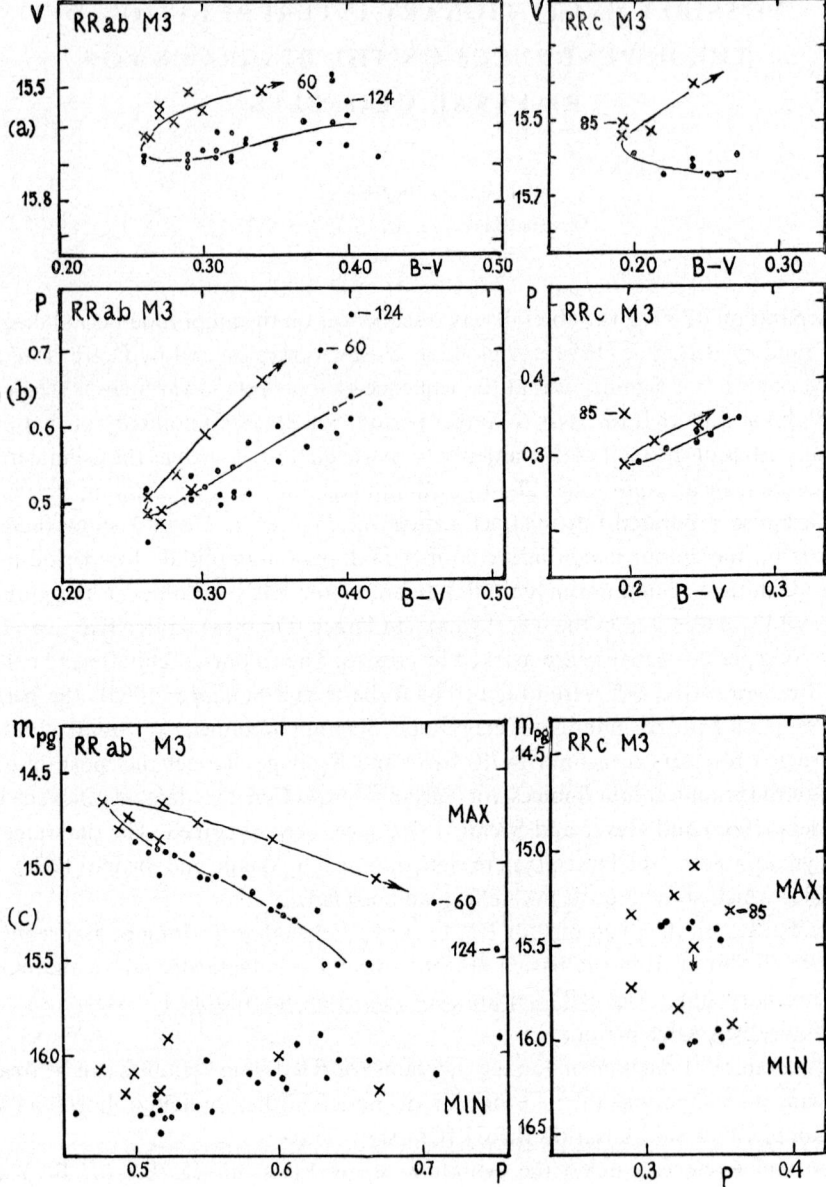

Fig. 1. The colour-magnitude (a), colour-period (b), period-maximum magnitude, minimum magnitude (c) diagrams for RR_{ab} and RR_c variables in the globular cluster M3. Solid lines are possible empirical tracks of variables.

imately constant. If pulsations are stimulated in the superficial layers of the star, the period of these pulsations will decrease.

On the $P - \log T_e$ diagram a variable will move approximately along the temperature sequence of models calculated by Christy for fixed mass, luminosity, and chemical composition (Figure 5). On the $A-P$ diagram a variable will shift along the sequence of models with the same fixed parameters (Figure 6). The group of variables of approximately equal mass at this stage of evolution will form the short period sequence.

According to the calculations, after relative exhaustion of the helium content in the core (to 30% of the initial content) the evolutionary changes undergo a revision. They lead to an increase of the luminosity and to an expansion of the envelope of the star. After the turn-off point of the evolutionary track the star evolves upwards and to the

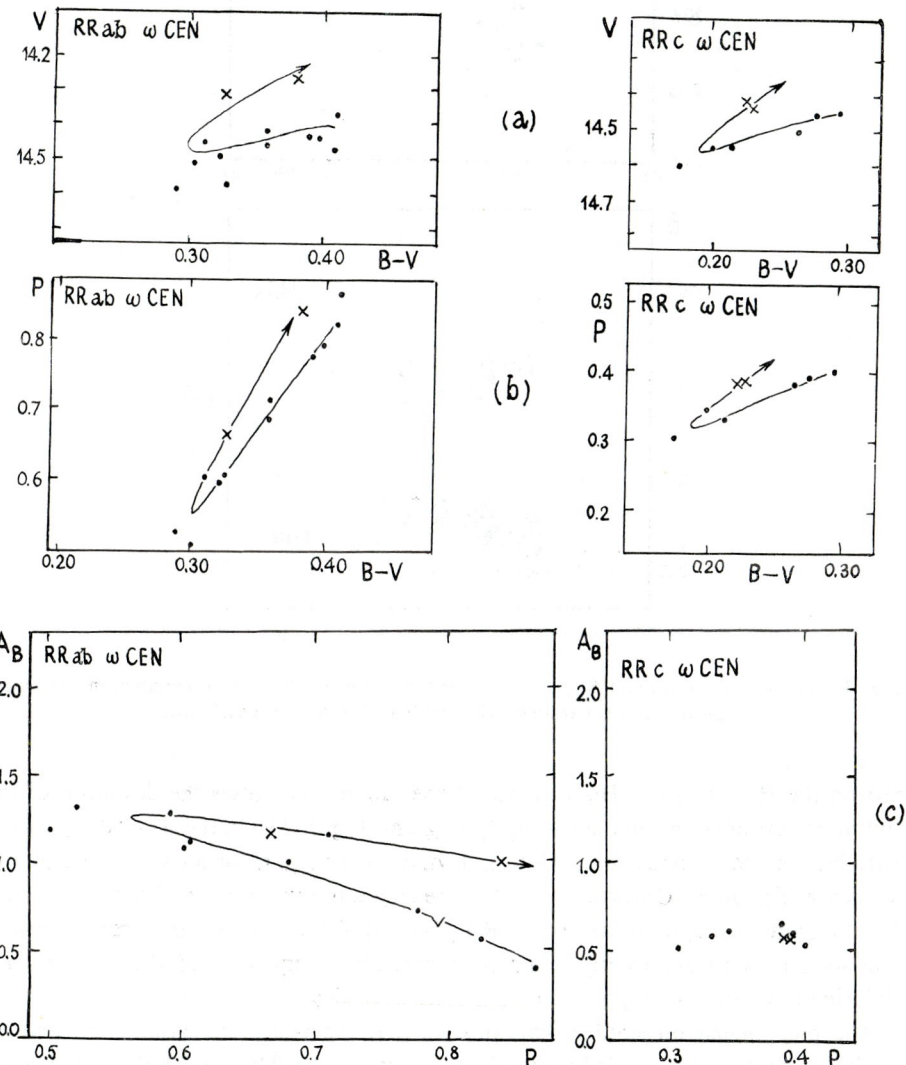

Fig. 2. The colour-magnitude (a), colour-period (b), period-amplitude (c) diagrams for RR_{ab} and RR_c variables in the globular cluster ω Cen. Solid lines are possible empirical tracks of variables. The stars known to show the Blazhko-effect are excluded.

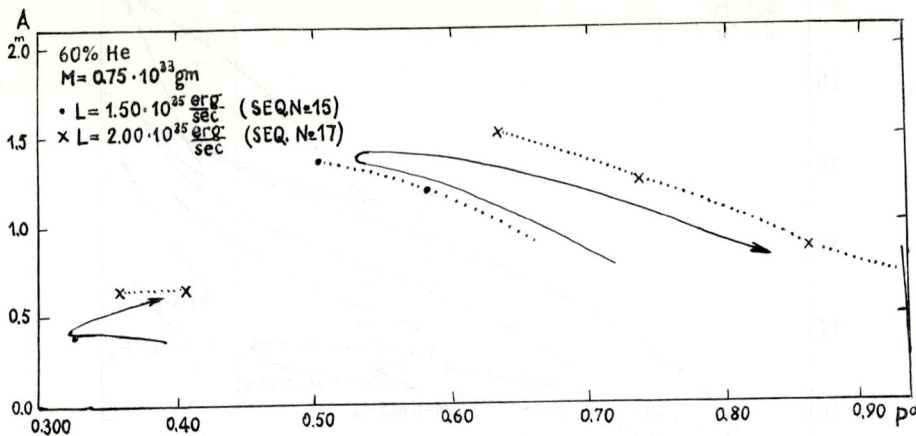

Fig. 6. Schematic representation of sequences in the period amplitude diagram. Dotted lines are the sequences of Christy.

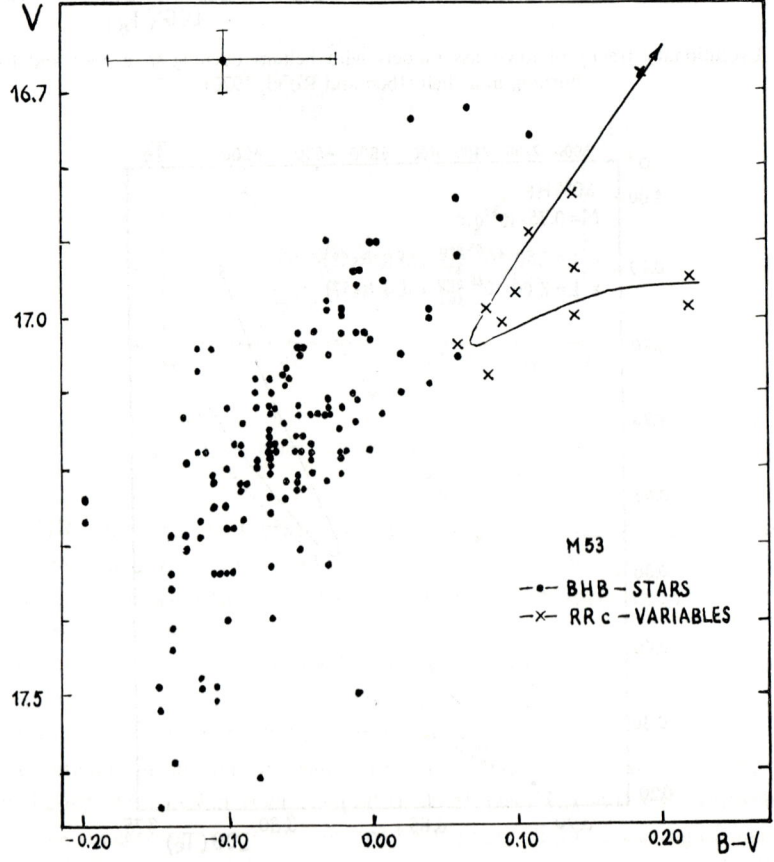

Fig. 7. The colour-magnitude diagram for BHB stars and RR_c variables in the globular cluster M53. The observed effect is of order of observational errors.

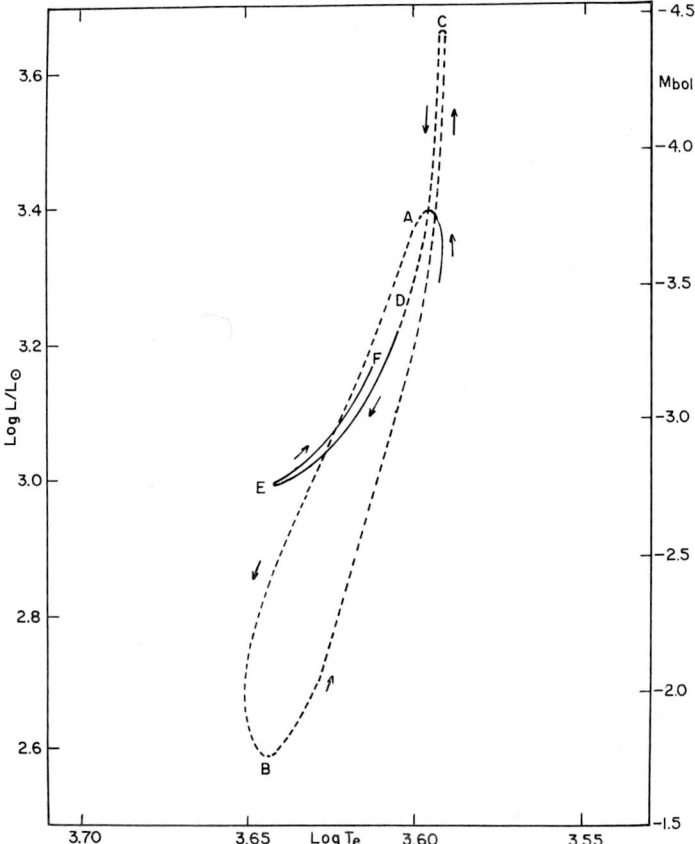

Fig 3. Track in the HR diagram during the active phase of cycle B. Dashed curve (A-B-C) represents a relatively fast phase immediately following the main peak in the helium burning. Solid curve (D-E-F) indicates the relatively slow drop in the surface luminosity at the end of the active phase. Times in years between designated points are: $t_{AB} = 500$, $t_{BC} = 740$, $t_{CD} = 7100$, $t_{DE} = 25\,000$, $t_{EF} = 34\,000$.

cycles which this star will undergo before it evolves off the asymptotic branch. These advanced cycles will be denoted by A, B, and C.

Figures 2 and 3 show the evolutionary track for the active phases of cycles A and B respectively. The solid line in Figure 2 represents the track during the rapid drop in L immediately following the main helium-burning peak, at which time the envelope mass was 0.027 $M_\odot$. The dashed line shows the corresponding phase for a model with the same total mass and envelope mass, but with the envelope composition (0.600; 0.001). Since no appreciable departure from the asymptotic branch was found at this phase for the case $X = 0.732$, the calculations for cycle A were terminated soon after the rapid drop in L. During the active phase of cycle B the envelope mass was 0.019 $M_\odot$. Computations were carried out only for the composition (0.732; 0.001), and the entire active phase was followed. Relatively fast and slow stages are indicated in Figure 3 by

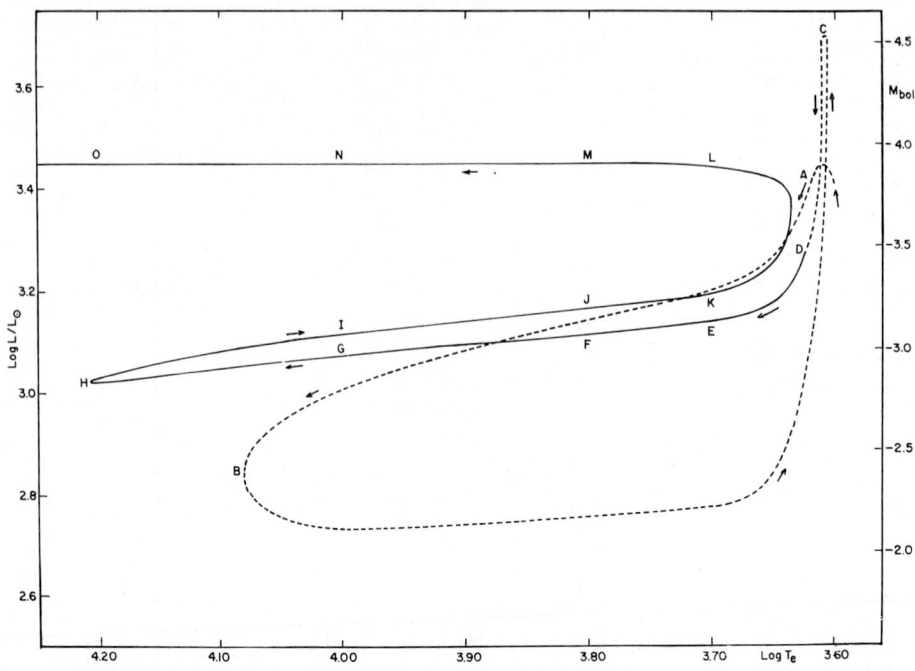

Fig. 4. Track in the HR diagram for the active phase of cycle C. Slow and fast phases are designated by solid and dashed lines respectively. The fast loop following the main peak in the helium burning is given by the segment of dashed curve from A to C. The slow loop at the end of the active phase is given by the segment of solid curve from D to K. Times in years between designated points are: $t_{AB} = 290$, $t_{BC} = 850$, $t_{CD} = 8650$, $t_{DE} = 7000$, $t_{EF} = 1730$, $t_{FG} = 3100$, $t_{GH} = 10700$, $t_{HI} = 16500$, $t_{IJ} = 7500$, $t_{JK} = 5800$, $t_{KL} = 170000$, $t_{LM} = 19300$, $t_{MN} = 13300$, $t_{NO} = 11900$.

the dashed and solid lines respectively. As was found for cycle A, the maximum departure from the asymptotic branch is quite small.

For the active phase of cycle C the envelope mass was 0.011 $M_\odot$. The calculations were made for the composition (0.732; 0.001). The evolutionary track is shown in Figure 4. The principal features of the track are a fast loop (points A to C) lasting 380 years, and a slow loop (points D to K) lasting 82000 yr. In the quiescent phase following cycle C the star evolved off the asymptotic branch (points K to O).

5. Discussion

These results indicate that significant departures from the asymptotic branch resulting from helium shell flashes will occur only when the outward progression of the hydrogen-burning shell has reduced the envelope mass to quite low values. For $(X; Z) = (0.732; 0.001)$ this mass must be less than about 0.02 $M_\odot$, while for (0.600; 0.001) it may be as large as 0.03 $M_\odot$. The requirement of small envelope mass means that loops will occur only for the last few relaxation cycles on the asymptotic branch. The precise number of cycles which produce loops remains somewhat uncertain, since the suppression of the early cycles introduces some arbitrariness into the models. This is because the exact time at which a flash occurs and therefore the envelope mass at that time depend on when the transition from 'flash suppressed' to 'flash permitting' computations is made. For the case (0.732; 0.001) an estimate of 1 or 2 loop-producing cycles is perhaps not unreasonable. A similar result should hold in the case of more massive stars when the envelope mass has been reduced to a value sufficiently small for the occurrence of loops.

The amount of time which the star spends in the instability strip is rather difficult to estimate because of the arbitrariness mentioned above, and because of uncertainties in the position of the edges of the instability strip. For the 0.60 $M_\odot$ star, lower and upper limits for this time may be crudely taken as 50000 and 200000 yr. These estimates include crossings of the instability strip both during loops, and during the final evolution off the asymptotic branch. If the beginning of the asymptotic-branch phase is taken to be the time when core convection ceases following the horizontal-branch stage, then these times represent about 0.2 to 0.8% of the asymptotic-branch lifetime of the star.

The present results predict rates of period change for the slow-loop and post-asymptotic-branch crossings of the instability strip that are about an order of magnitude slower than those found by SH and therefore agree better with observation. Period changes during a fast-loop crossing might be observable, but the probability of observing a star in such a relatively rapid stage is small.

6. Summary

The results of this investigation confirm in general the preliminary results of SH. The principal difference is that SH found that the tracks of asymptotic-branch stars of low mass describe a large number of fast loops, whereas the present results suggest the occurrence of a small number of slow loops. These loops, together with the final evolution off the asymptotic branch, provide a possible explanation for the occurrence in globular clusters of Cepheids with periods greater than about 8 days. The Cepheids with shorter periods can perhaps be identified as stars evolving off the horizontal branch, as has been suggested by Kraft (1972).

Acknowledgements

The author wishes to thank Pierre Demarque and Allen Sweigart for their interest and advice throughout the course of this research. The support of grant NSF GP 21345 from the National Science Foundation is gratefully acknowledged.

References

Cox, A. N. and Stewart, J. N.: 1970, *Astrophys. J. Suppl.* **19**, 261.
Kraft, R. P.: 1972, in A. G. Davis Philip (ed.), *The Evolution of Population II Stars*, Dudley Observatory Report No. 4, Albany, p. 69.

Schwarzschild, M.: 1971, private communication.
Schwarzschild, M. and Härm, R.: 1965 *Astrophys. J.* **142**, 855.
Schwarzschild, M. and Härm, R.: 1967, *Astrophys. J.* **150**, 961.
Schwarzschild, M. and Härm, R.: 1970, *Astrophys. J.* **160**, 341.
Sweigart, A. V.: 1971, *Astrophys. J.* **168**, 79.
Sweigart, A. V.: 1972, *Bull. Am. Astron. Soc.* **4**, 202.

DISCUSSION

Schwarzschild: I would like to emphasise two points of Dr Mengel's remarks. First, because Dr Mengel treats the envelope of his stars enormously more carefully than we did in Princeton I believe that he is right in saying that the fast loops in the early cycles probably do not occur to any significant extent, contrary to our earlier and rougher Princeton results. Second, unpublished rough Princeton computations have also given the slow loops in the advanced cycles; this I feel adds some weight to Dr Mengel's discovery of these slow loops and their possible role in forming the long-period group of Population II Cepheids.

Lloyd Evans: There are type *II* Cepheids of the short period ($P \sim 2^d$) group in the disk population which show enhanced abundance of C and, possibly, s-process elements.

Demers: How could we explain the lack of Population II Cepheids with $P > 3^d$ in dwarf spheroidal galaxies? Differences in chemical composition? Earlier evolutionary stage of these galaxies?

Mengel: Perhaps the masses of asymptotic branch stars in these systems are a bit higher. The slow loops would be somewhat speeded up in that case, and one might not see stars in this phase. There is also the possibility that the envelope might be lost before the slow loop stage is reached for a star with $M > 0.60 \, M_\odot$. If that happens, the slow loops may not occur at all, as Prof. Schwarzschild has pointed out.

Dickens: Can you say anything from your models about the possibility of mixing from the hydrogen burning shell into the atmosphere?

Mengel: This has been looked into by Allen Sweigart and myself for a $0.70 \, M_\odot$ asymptotic branch star. At one point following a helium shell flash the surface convective zone becomes very deep. (This occurs during the rapid rise in the surface luminosity following the burning peak of a helium shell flash.) Surface convection reaches in to the top of the hydrogen-burning shell, but does not penetrate very far into it, even in this favorable case. The hydrogen shell is so tightly bound that the pressure difference from its top to its bottom amounts to many orders of magnitude. Since surface convection did not reach deeply into the shell, we were unable to mix any CNO-processed material to the surface.

SEMICONVECTION AND THE RR LYRAE VARIABLES

A. V. SWEIGART and P. DEMARQUE

Yale University Observatory, U.S.A.

1. Introduction

Theoretical computations (Hoyle and Schwarzschild, 1955; Faulkner, 1966; Iben and Rood, 1970; Demarque and Mengel, 1971a, b) have identified the horizontal-branch stars in globular clusters with the evolution phase in which helium burns within a convective core and hydrogen burns in a shell outside the convective core. Most computations for such double-energy-source models have indicated that the evolution proceeds smoothly on a nuclear time scale during the horizontal-branch phase, leading to small predicted rates of change in the RR Lyrae pulsation period (Iben and Rood, 1970). Sweigart and Demarque (1972) have recently considered the effects of semiconvection on the horizontal-branch evolution of typical Population II stars and have suggested that changes in the composition distribution within the core may occur on a time scale considerably shorter than the nuclear time scale during the phase immediately preceding core-helium exhaustion. It has been found that the composition distribution generated by the growth of a semiconvective zone in the layers surrounding the convective core can become unstable when Y_c, the helium abundance within the convective core, decreases below roughly 0.12. The changes in the internal structure caused by this instability result in relatively rapid movement of the models in the HR diagram and consequently produce large predicted rates of change in the RR Lyrae pulsation period. The possibility that RR Lyrae period changes may be associated with the behavior of the semiconvective zone has been previously suggested by Schwarzschild (1970). A similar instability may occur in the late core-hydrogen burning phase for stars around 10 $M_\odot$. Percy (1970) has noted the coincidence of β Cephei stars with stellar models containing semiconvective zones. It is tempting to suggest that such an instability in the semiconvective zone could also be related to the β Cephei phenomenon.

The principal purposes of the remaining sections are as follows: (1) to discuss the horizontal-branch evolution of a 0.65 $M_\odot$ Population II star (Section 2), (2) to describe the instability in the core-composition distribution (Section 3) and (3) to illustrate the rates of period change which this instability produces for this star (Section 4).

2. Horizontal-Branch Evolution of a 0.65 $M_\odot$ Star

A 0.65 $M_\odot$ star with the envelope composition parameters $X=0.732$ and $Z=0.001$ has been evolved from the initial horizontal-branch phase to the point of helium exhaustion in the core (Sweigart and Demarque, 1972). The evolutionary track, presented in Figure 1, was computed for an initial core mass of 0.468 $M_\odot$.

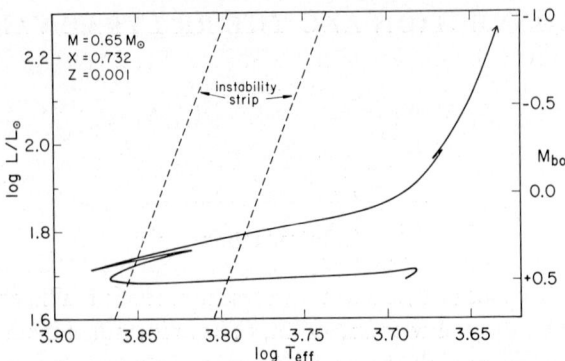

Fig. 1. The horizontal-branch track for a 0.65 $M_\odot$ star with the envelope composition parameters $X = 0.732$ and $Z = 0.001$. Approximate limits for the instability strip taken from Iben and Rood (1970, Figure 7) are indicated by the dashed lines.

The model construction has been performed with generally the same numerical techniques as those adopted by Sweigart and Demarque (1972). Radiative opacities were obtained from Cox and Stewart (1970), and a mixing-length equal to 1 pressure scale height was used in treating the surface convection zone. The reaction $^{12}C(\alpha, \gamma)^{16}O$ was not included in determining the helium burning rate. Lauterborn et al. (1972) have shown that the basic features of the horizontal-branch evolution are insensitive to the omission of this reaction.

The investigations of Schwarzschild and Härm (1969), Paczyński (1970), Castellani et al. (1971b) and Demarque and Mengel (1972) have demonstrated that stars with masses less than several solar masses will form a semiconvective zone in the region adjacent to the convective core during the phase of core-helium burning. The method proposed by Robertson and Faulkner (1972) has been employed to treat the occurrence of semiconvection in the present 0.65 $M_\odot$ star. Failure to incorporate such a treatment of semiconvection can lead to a substantial underestimate of the lifetime of the horizontal-branch evolution and of the length of the horizontal-branch track in terms of log T_{eff} (Demarque and Mengel, 1972; Sweigart and Demarque, 1972).

During the initial horizontal-branch evolution of this 0.65 $M_\odot$ star the mass content of the convective core grows as a result of convective overshooting (Castellani et al., 1971a). The reason for this growth is the necessity to fulfill the condition of convective neutrality just inside the convective-core edge (Schwarzschild, 1958). This stage of convective overshooting lasts for 17×10^6 yr during which Y_c is reduced to a value of 0.82 and during which the convective-core mass increases from 0.105 to 0.147 $M_\odot$. The interior composition distribution in the models of this stage therefore consists of a somewhat helium-depleted convective core which is separated by a sharp composition discontinuity from an exterior region of nearly pure helium in radiative equilibrium.

For $Y_c < 0.82$ continued expansion of the convective core is no longer capable of preserving neutrality on the inner side of the convective-core edge due to a characteristic turning up of the radiative gradient V_{rad} ($=d\log T/d\log P$) with increasing values of M_r (Castellani et al., 1971b; Demarque and Mengel, 1972). The star resolves this dilemma by creating a semiconvective zone in the layers immediately outside the convective core. In this semiconvective zone the helium abundance Y varies smoothly by the amount needed to enforce the requirement of convective neutrality.

During the evolution to the blue along the main part of the horizontal-branch track the mass in the semiconvective zone increases while the mass in the convective core remains almost constant at a value of 0.15 $M_\odot$. At the bluest point of the main horizontal-branch track ($\log T_{eff}=3.866$, $\log L/L_\odot=1.692$) Y_c has decreased to 0.28. The semiconvective zone reaches its maximum extent shortly after the star begins its initial return to the red. The interior composition distribution at this stage is depicted in Figure 2. The semiconvective zone in this figure encompasses the mass between $M_r=0.148$ and $0.256\ M_\odot$ and thus contains about $0.11\ M_\odot$. Following this stage the

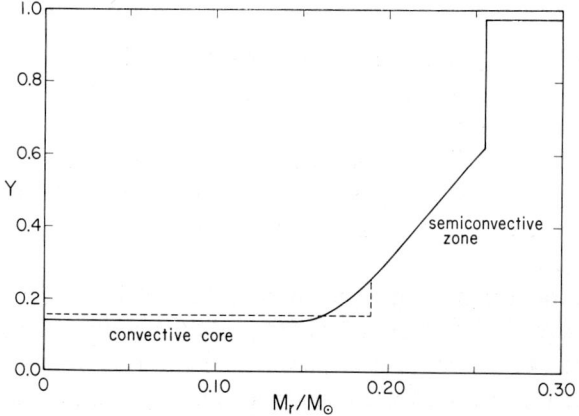

Fig. 2. The composition distribution within the core of a Population II horizontal-branch star of 0.65 $M_\odot$ at the stage of maximum extent of the semiconvective zone. The dashed line represents qualitatively the modification in the composition profile produced by overshooting at the convective-core edge.

outer edge of the semiconvective zone recedes, leaving behind a region of varying helium abundance that is stable against convection by a small amount. When Y_c reaches a value of 0.12, the composition distribution within the core undergoes a strong perturbation caused by the instability described in Section 3. It is this perturbation which leads to the abrupt blueward shift of the track in Figure 1 at $\log T_{eff}=3.818$.

3. Instability of the Composition Distribution Within the Core

One might normally expect that, subsequent to the stage shown in Figure 2, there would be a gradual depletion of the remaining helium supply within the convective core. Some reasons exist, however, for suspecting that the evolution just prior to core-helium exhaustion might not proceed in such a smooth fashion (Sweigart and

Demarque, 1972). Assume that the mass content of the convective core increases somewhat because of convective overshooting and that consequently Y_c is enriched by an amount δY. Even a small positive perturbation δY will enhance the rate of energy generation from the triple-α process by a relatively large amount if Y_c is sufficiently small. A significant enhancement in the helium burning rate might then raise the radiative gradient ∇_{rad} in the neighborhood of the convective-core edge, thereby maintaining the convective overshooting. Such a readjustment of the models would thus represent an instability in the core-composition distribution that could permit a considerable growth in the convective-core mass. The convective overshooting needed to initiate this instability is provided by the increase in the radiative opacity that occurs with the depletion of Y_c (Castellani et al., 1971a). The driving mechanism is due to the progressively greater effect which a perturbation δY of fixed magnitude has on the total helium-burning rate as Y_c is lowered. These general arguments have been confirmed by a stability analysis and by model computations, both of which we now discuss.

To perform a stability analysis of the core-composition distribution requires a knowledge of the way a model readjusts in response to a perturbation $\delta Y(M_r)$ in the helium abundance. An expansion of the convective core due to overshooting will modify the composition distribution in the manner indicated qualitatively by the dashed curve in Figure 2. At each point we define $\delta Y(M_r)$ to be the difference between the dashed and solid curves of Figure 2. Outside the point of maximum overshooting $\delta Y(M_r)$ is set equal to 0.

The set of equations governing the readjustment is obtained by expanding the four stellar structure equations to first order in terms of the perturbations δr, δL_r, δP, δT and δY. The basic procedure is to impose a prescribed perturbation $\delta Y(M_r)$ on a model and to determine from these linearized equations the resulting perturbations in the remaining physical variables. This solution immediately yields the perturbation in the ratio of the radiative to adiabatic gradients, i.e. $\delta(\nabla_{rad}/\nabla_{ad})$.

An essential ingredient for choosing the proper solution is an estimate for the time scale during which the overshooting used in specifying $\delta Y(M_r)$ takes place, since the solution depends directly on the assumed time scale. Castellani et al. (1971a) have investigated the efficiency of overshooting on the basis of the mixing-length theory. They provide a convenient relationship for the amount of overshooting which involves the degree of superadiabaticity and the change in composition at a convective boundary.

Stability of the core-composition distribution is defined by the requirement that $\delta(\nabla_{rad}/\nabla_{ad})$ be negative in the region surrounding the convective-core edge. Such a readjustment would automatically prohibit an abrupt growth of the convective core by suppressing the overshooting. The condition for instability is consequently that $\delta(\nabla_{rad}/\nabla_{ad})$ be positive in this region. Employing this criterion, we find that the models are stable roughly for $Y_c \gtrsim 0.12$ and unstable for $Y_c \lesssim 0.12$.

The behavior predicted by the above stability test is displayed by the models for the 0.65 $M_\odot$ star. In computing these models the amount of overshooting which occurs at the convective-core edge over a time step has been determined from the relationship given by Castellani et al. (1971a, Equation (5′)) and is thus consistent with physical conditions prevailing at the convective-core edge.

Figure 3 gives the time dependence of Y_c from the 0.65 $M_\odot$ sequence. The value of Y_c increases from 0.12 to 0.20 over an interval of approximately 10^6 yr due to the movement of the convective-core edge out to the mass point $M_r = 0.256\ M_\odot$. The radiative gradient in the expanded convective core exhibits the characteristic mini-

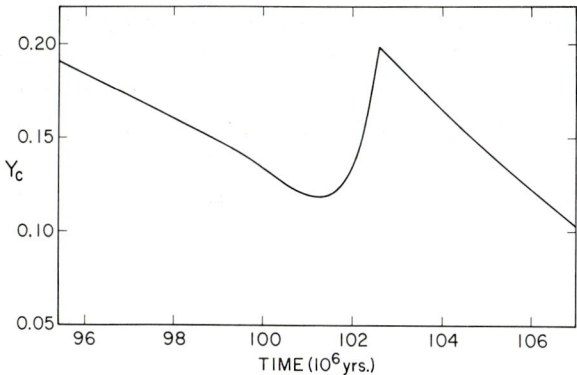

Fig. 3. Time dependence of Y_c, the helium abundance within the convective core, prior to core-helium exhaustion in a Population II horizontal-branch star of 0.65 $M_\odot$. The time scale gives the time elapsed since the initial horizontal-branch phase.

mum referred to in Section 2 (Demarque and Mengel, 1972). The increase in Y_c is terminated when this minimum falls below the adiabatic gradient. The basic features of this behavior resemble those of the 'core breathing' phenomenon discussed by Castellani et al. (1971b) and Demarque and Mengel (1972) except that in the present case the nuclear time scale exceeds the overshooting time scale. During the Y_c increase the overshooting time scale governs the rate of Y_c enrichment.

4. Predicted Rates of Change in the RR Lyrae Pulsation Period

As demonstrated in Section 3, Y_c does not decrease monotonically near the end of the horizontal-branch phase but rather undergoes at least one period of relatively rapid enrichment. Such a variation in Y_c represents a significant change in the internal structure and thus strongly influences the shape of the theoretical track in the HR diagram. The increase in Y_c around the time of 102×10^6 yr in Figure 3 forces the 0.65 $M_\odot$ models to reverse their redward movement in Figure 1 at $\log T_{\text{eff}} = 3.818$ and $\log L/L_\odot = 1.758$. The subsequent movement to the blue continues for 1.5×10^6 yr until the $\log T_{\text{eff}}$ value reaches a maximum of 3.878. This maximum occurs once Y_c stops increasing and is followed by yet another abrupt change in the direction of the track. The models then evolve redward through the instability strip on a nuclear time scale and approach the asymptotic branch. By the end of the track shown in Figure 1,

DISCUSSION

Schwarzschild: The observations I think force us to look for some kind of instability or disorderliness in the horizontal branch evolution. The data on period changes seem rather convincing, I believe, in showing that superimposed on a barely detectable evolutionary period drift there exists for most RR Lyrae variables seemingly random period changes (possibly abrupt, see for example, Prager, *Harvard Bulletin* No. 911, 1939). Dr Sweigart has shown, I think quite persuasively, that there can exist an instability in the extent of the convective core which will seriously perturb the period. I guess that both Dr Sweigart and I regret greatly that his new instability seems to occur at most once in the instability strip, while the observations seem to require a disturbance (with more or less random signs) roughly once every 100 years.

Sweigart: Three difficulties exist in attempting to explain the random period fluctuations with a time scale of once every 100 years on the basis of the present instability of the composition distribution within the core of horizontal-branch stars. First, our 0.65 $\mathfrak{M}_\odot$ star became unstable only once while within the instability strip. Secondly, the characteristic time scale of this instability, governed by the efficiency of overshooting at the convective-core edge, far exceeds the time scale of the period fluctuations. Thirdly, this instability only gives rise to large negative rates of period change. For these reasons Dr Schwarzschild is quite correct in emphasizing the need to search for other sources of disorderliness in horizontal-branch stars. It is not implausible that such disorderliness may be associated with the behavior of the semiconvective zone. The present instability may, however, offer a possible explanation for RR Lyrae variables which show large negative rates of period change in addition to any random period fluctuations.

Clement: Why doesn't this chemical instability repeat?

Sweigart: A second period of instability in our 0.65 $\mathfrak{M}_\odot$ star was actually found when the helium abundance within the convective core had been reduced to 0.02. At that time, however, this star was evolving up the asymptotic branch and therefore was no longer of interest in regard to the RR Lyrae variables. The repetition of this instability was quite mild, since the mass content of the convective core did not greatly increase.

THE HELIUM ABUNDANCE IN THE ENVELOPES OF THE BLUEST RR LYRAE STARS IN GLOBULAR CLUSTERS AND DEPENDENCE OF GLOBULAR CLUSTER VARIABLE STAR PROPERTIES ON CHEMICAL COMPOSITION

A. V. MIRONOV

Tien-Shan High Mountain Observatory, Sternberg State Astronomical Institute, U.S.S.R.

Abstract. The helium abundance Y in the envelopes of RR Lyrae stars in globular clusters has been estimated. The values of Y range from 0.07 to 0.59. The properties of variable stars in globular clusters of two types distinguished by a type of dependence of the horizontal branch form on the chemical composition are compared. The clusters of type I are shown to be on the average poorer in RR Lyrae stars than those of type II. The RR Lyrae stars in type I clusters are on the average brighter by 0.1 mag. It is found that as Y increases, the cluster richness in W Virginis variables decreases.

It has long been known that a number of properties of globular cluster variable stars depend on chemical abundance.

In the present report we shall discuss some of these properties in connection with our discovery of the separation of globular clusters into two groups distinguished by a type of dependence of the horizontal branch form on chemical abundance.

Recently Kukarkin has developed a system of metallicity indexes (IM) based on many determinations by other authors (Kukarkin and Russev, 1972) of metal abundance in the atmospheres of globular cluster stars. Apparently the metallicity indexes most fully and objectively reflect our modern knowledge of metal abundance in globular cluster stars.

Sandage (1969), guided by Christy's (1966) theoretical calculations, suggested a method of determination of helium abundance from the $(B-V)$ colour of the blue edge of the RR Lyrae strip on the Hertzsprung-Russell diagram. Sandage (1969, 1970) applied his method for obtaining the helium abundance in horizontal branch stars of the globular clusters M3, M13, M15, and M92. But at the present time the diagrams for 41 globular clusters have been published. For 27 of them the colour of the blue edge of the RR Lyrae strip is determined with sufficient confidence. We have somewhat altered Sandage's method, having applied Newell's (1970) colour-effective temperature relation for blue horizontal branch stars and having taken into account the results of new calculations of pulsation models made by Iben and Huchra (1970). Using this altered method, we have determined the helium abundance Y in the envelopes of the bluest RR Lyrae stars in 27 globular clusters (Mironov, 1971, 1972). The values of Y obtained range from 0.07 to 0.59.

The comparison of the quantities IM and Y and their product $IM \cdot Y$ with the parameters of the horizontal branch form shows that all globular clusters can be divided into two groups. To quantitatively describe the horizontal branch form we consider two new parameters. The first of them is the number of blue stars (i.e. stars to the left of the RR Lyrae strip expressed as a fraction of the total number of non-variable horizontal branch stars. The second parameter is the difference in $(B-V)$ between the red end of the horizontal branch and that point on the horizontal branch at which the greatest concentration of stars occur. The former point is designated as $(B-V)_2$, the latter one as $(B-V)_1$.

The resulting separation of globular clusters into two groups according to the type of dependence of the horizontal branch form on the abundance is illustrated by Figures 1 and 2, where the type I clusters are indicated by crosses and the type II clusters by filled circles.

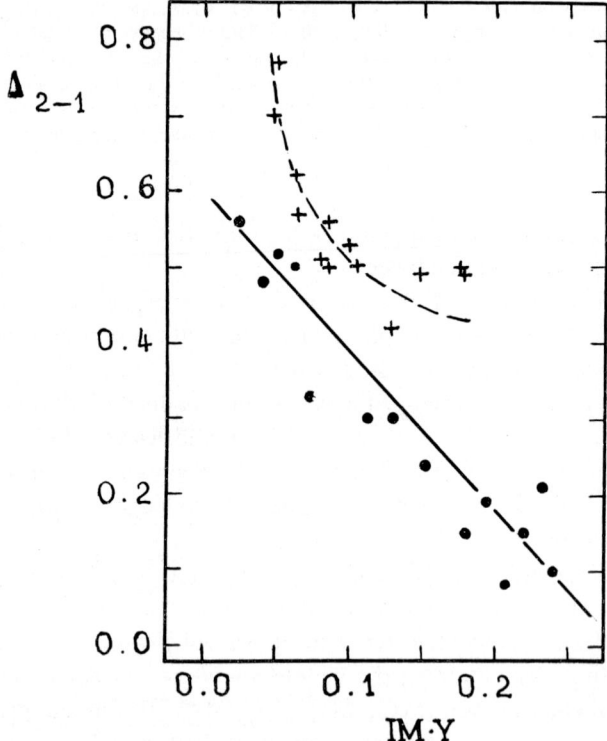

Fig. 1. The dependence of the colour difference between the red end of the horizontal branch and the point of maximum star concentration in the HB as a function of chemical abundance. Crosses indicate type I cluster, circles type II clusters.

The other properties of the variable stars in these two classes of clusters also differ, which can be seen from Table I, where they are listed with the abundance parameters.

The first column of this table contains the designation of clusters. The abundance parameters IM, Y and $IM \cdot Y$ are given in the next three columns. The type I clusters have a rather small range of metallicity indexes, and their metal abundance is on the average lower than that of the type II clusters.

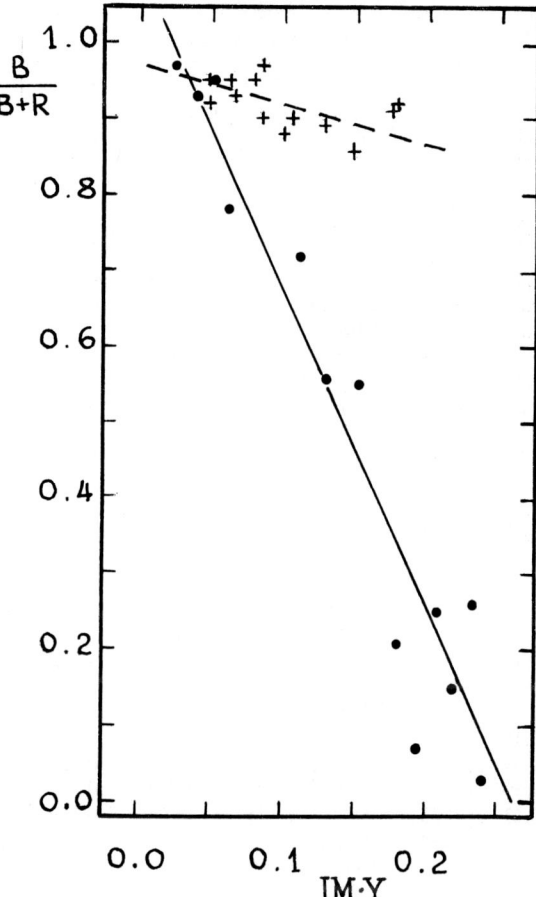

Fig. 2. The number of HB stars bluer than RR Lyr stars (expressed as a fraction of all non-variable HB stars) vs chemical abundance. Symbols as in Figure 1.

The number N_{RR} of RR Lyrae stars in the clusters under consideration are given in the fifth column, and the sixth column contains the same numbers corrected for the richness of a cluster N'_{RR}. Such corrections are quite necessary when comparing the number of variables in various clusters, as all other things being equal, the probability of finding a star of some type is proportional to the total number of stars in a cluster, i.e. to cluster richness. We made this correction by using a factor that is inversely proportional to the absolute cluster luminosity, the richness of the absolutely brightest cluster, ω Cen, being unity. It is obvious that type I clusters are essentially poorer in RR Lyrae stars than the type II clusters. As Figure 3 shows, the relative quantity of RR Lyrae variables in both types of clusters increases at first, and then begins decreasing, the type I cluster curve always staying lower than the type II cluster curve. This dependence can be easily explained if we take into account that with the increasing of $IM \cdot Y$ the $(B-V)_1$ point moves from the blue area into the red one. The

TABLE I
Properties of globular clusters

Clusters NGC name (1)		IM (2)	Y (3)	IM·Y (4)	N_{RR} (5)	N'_{RR} (6)	P_{tr} (7)	M_{RR}^{bol} (8)	N_{cw} (9)	N'_{cw} (10)
Type I Clusters										
4147		0.35	0.42	0.147	9	320.8	0.504	+0.78	?	–
5024	M53	0.30	0.59	0.177:	42	210.4	0.529	+0.69	1	5.01
5139	ω Cen	0.33	0.26	0.086	142	142.0	0.518	+0.73	7	7.00
5466		0.27	0.47:	0.127:	18	525.2	0.542	+0.65	?	–
5897		0.31	0.21	0.065	6	87.5	–	–	?	–
6205	M13	0.36	0.24	0.086	3	16.6	–	–	3	16.65
6218	M12	0.38	0.26	0.099	0	0	–	–	1	12.94
6254	M10	0.36	0.13	0.047	0	0	–	–	2	23.40
6397		0.33	0.15:	0.050:	0	0	–	–	?	–
6752		0.33	0.53	0.175	0	0	–	–	?	–
7089	M2	0.31	0.26	0.081	17	53.7	0.520	+0.72	4	12.66
7492		0.31	0.34	0.105	0	0	–	–	?	–
mean		0.33	0.32	0.104	20	113.0	0.520	+0.71	3	12.94
Type II Clusters										
104	47 Tuc	0.57	0.42	0.239	2	4.7	–	–	?	–
362		0.43	–	–	7	36.7	0.474	+0.89	?	–
4833		0.34	0.07	0.024	6	29.3	–	–	?	–
5053		0.26	0.28	0.073	10	506.2	0.576	+0.54	?	–
5272	M3	0.37	0.35	0.130	201	910.5	0.490	+0.83	1	4.53
5904	M5	0.36	0.31	0.112	99	514.8	0.500	+0.80	2	10.40
6121	M4	0.49	0.31	0.152	41	880.7	0.433	+1.06	?	–
6171		0.50	0.39	0.195	22	522.9	0.438	+1.03	?	–
6341	M92	0.23	0.22	0.051	15	109.6	0.560	+0.59	?	–
6362		0.43	0.59:	0.254:	15	310.5	0.460	+0.94	?	–
6656	M22	0.31	0.13	0.040	19	129.0	0.547	+0.63	3	20.37
6712		0.50	0.44	0.220	10	185.4	0.481	+0.87	?	–
6723		0.53	0.44	0.233	19	419.5	0.444	+1.01	?	–
7006		0.39	0.53:	0.207:	74	818.4	0.490	+0.83	?	–
7078	M15	0.24	0.26	0.062	93	356.9	0.575	+0.54	2	7.46
mean		0.40	0.33	0.135	42	381.7	0.498	+0.81	2	10.69

maximum number of RR Lyrae stars naturally occurs when this point is situated in the middle of the instability strip, the colour of which varies comparatively little among clusters.

The seventh column lists the values of a critical transition period P_{tr} from the pulsations of the *ab* type to the *c* type. According to Christy's (1966) investigation this value is related to the luminosity of the RR Lyrae stars by the equation

$$M_{RR}^{bol} = 0.46 - 4.17 \log P_{tr}.$$

Using this formula, we find that the RR Lyrae variables in the type I clusters are brighter by a 0.1 mag. than in the type II clusters.

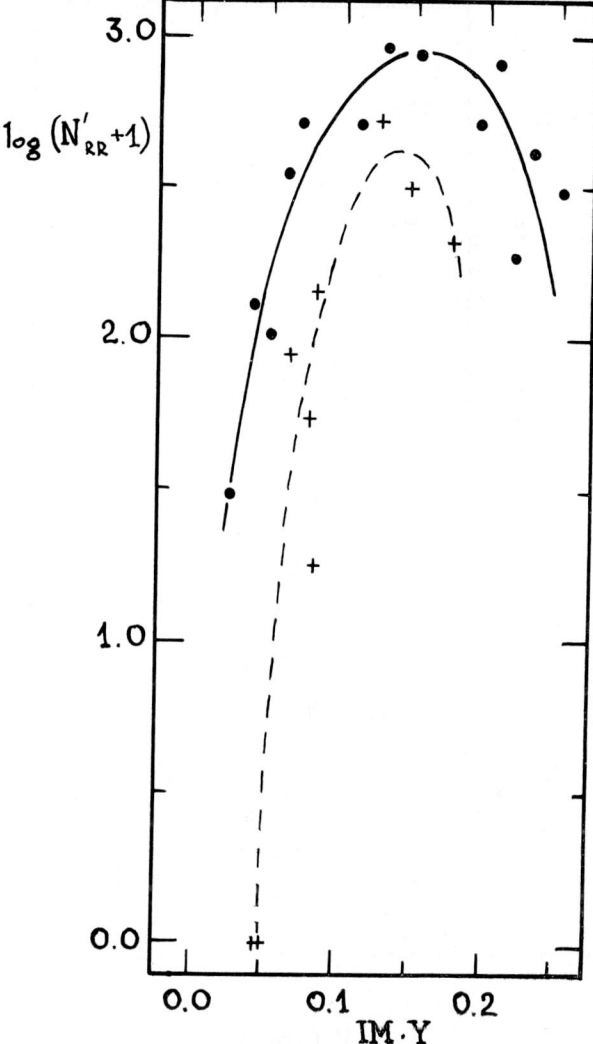

Fig. 3. Relative richness of RR Lyrae population in clusters as a function of chemical abundance. Symbols as in Figure 1.

The ninth and tenth columns are analogous to the fifth and sixth columns, respectively, but relate now to W Virginis variables. Because of the small number of these variables in individual clusters, no reliable conclusions can be drawn. Nevertheless our work confirms Wallerstein's (1970) conclusion that the W Vir variables do not occur in clusters of relatively high metallicity. Besides, the relation between the helium abundance Y and the value of N'_{cw} is interesting, as Figure 4 shows, N'_{cw} decreases as Y increases.

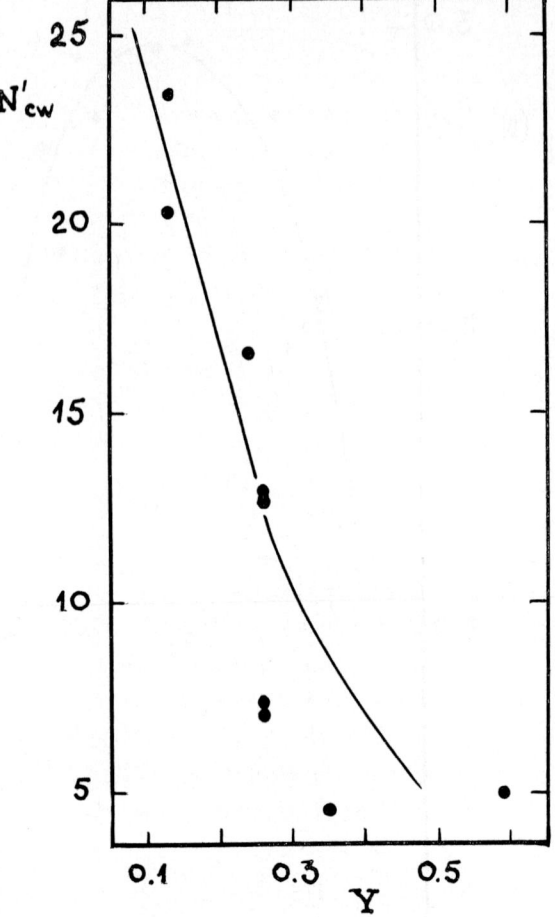

Fig. 4. The dependence of numbers of W Vir stars on helium abundance.

References

Christy, R. F.: 1966, *Astrophys. J.* **144**, 108.
Iben, I., Jr. and Huchra, J.: 1970, *Astrophys. J. Letters* **162**, L43.
Kukarkin, B. V. and Russev, R. M.: 1972, *Astron. Zh.* in press.
Mironov, A. V.: 1971. *Astron. Tsirk.* No. 667.
Mironov, A. V.: 1972, *Astron. Zh.* in press.
Newell, E. B.: 1970, *Astrophys. J.* **159**, 443.
Sandage, A.: 1969, *Astrophys. J.* **157**, 515.
Sandage, A.: 1970, *Astrophys. J.* **162**, 841.
Wallerstein, G.: 1970, *Astrophys. J.* **160**, 345.

ASTROPHYSICS AND SPACE SCIENCE LIBRARY

Edited by

J. E. Blamont, R. L. F. Boyd, L. Goldberg, C. de Jager, Z. Kopal, G. H. Ludwig, R. Lüst,
B. M. McCormac, H. E. Newell, L. I. Sedov, Z. Švestka, and W. de Graaff

1. C. de Jager (ed.), *The Solar Spectrum. Proceedings of the Symposium held at the University of Utrecht, 26–31 August, 1963.* 1965, XIV + 417 pp.
2. J. Ortner and H. Maseland (eds.), *Introduction to Solar Terrestrial Relations. Proceedings of the Summer School in Space Physics held in Alpbach, Austria, July 15–August 10, 1963 and Organized by the European Preparatory Commission for Space Research.* 1965, IX + 506 pp.
3. C. C. Chang and S. S. Huang (eds.), *Proceedings of the Plasma Space Science Symposium, Held at the Catholic University of America, Washington, D.C., June 11–14, 1963.* 1965, IX + 377 pp.
4. Zdeněk Kopal, *An Introduction to the Study of the Moon.* 1966, XII + 464 pp.
5. Billy M. McCormac (ed.), *Radiation Trapped in the Earth's Magnetic Field. Proceedings of the Advanced Study Institute, Held at the Chr. Michelsen Institute, Bergen, Norway, August 16–September 3, 1965.* 1966, XII + 901 pp.
6. A. B. Underhill, *The Early Type Stars.* 1966, XIII + 282 pp.
7. Jean Kovalevsky, *Introduction to Celestial Mechanics.* 1967, VIII + 427 pp.
8. Zdeněk Kopal and Constantine L. Goudas (eds.), *Measure of the Moon. Proceedings of the Second International Conference on Selenodesy and Lunar Topography held in the University of Manchester, England, May 30–June 4, 1966.* 1967, XVIII + 479 pp.
9. J. G. Emming (ed.), *Electromagnetic Radiation in Space. Proceedings of the Third ESRO Summer School in Space Physics, held in Alpbach, Austria, from 19 July to 13 August, 1965.* 1968, VIII + 307 pp.
10. R. L. Carovillano, John F. McClay, and Henry R. Radoski (eds.), *Physics of the Magnetosphere. Based upon the Proceedings of the Conference held at Boston College, June 19–28, 1967.* 1968, X + 686 pp.
11. Syun-Ichi Akasofu, *Polar and Magnetospheric Substorms.* 1968, XVIII + 280 pp.
12. Peter M. Millman (ed.), *Meteorite Research. Proceedings of a Symposium on Meteorite Research held in Vienna, Austria, 7–13 August, 1968.* 1969, XV + 941 pp.
13. Margherita Hack (ed.), *Mass Loss Stars. Proceedings of the Second Trieste Colloquium on Astrophysics, 12–17 September, 1968.* 1969, XII + 345 pp.
14. N. D'Angelo (ed.), *Low-Frequency Waves and Irregularities in the Ionosphere. Proceedings of the 2nd ESRIN-ESLAB Symposium, held in Frascati, Italy, 23–27 September, 1968.* 1969, VII + 218 pp.
15. G. A. Partel (ed.), *Space Engineering. Proceedings of the Second International Conference on Space Engineering, held at the Fondazione Giorgio Cini, Isola di San Giorgio, Venice, Italy, May 7–10, 1969.* 1970, XI + 728 pp.
16. S. Fred Singer (ed.), *Manned Laboratories in Space. Second International Orbital Laboratory Symposium.* 1969, XIII + 133 pp.
17. B. M. McCormac (ed.), *Particles and Fields in the Magnetosphere. Symposium Organized by the Summer Advanced Study Institute, held at the University of California, Santa Barbara, Calif., August 4–15, 1969.* 1970, XI + 450 pp.
18. Jean-Claude Pecker, *Experimental Astronomy.* 1970, X + 105 pp.
19. V. Manno and D. E. Page (eds.), *Intercorrelated Satellite Observations Related to Solar Events. Proceedings of the Third ESLAB/ESRIN Symposium held in Noordwijk, The Netherlands, September 16–19, 1969.* 1970, XVI + 627 pp.
20. L. Mansinha, D. E. Smylie and A. E. Beck, *Earthquake Displacement Fields and the Rotation of the Earth. A NATO Advanced Study Institute Conference Organized by the Department of Geophysics, University of Western Ontario, London, Canada, June 22–28, 1969.* 1970, XI + 308 pp.
21. Jean-Claude Pecker, *Space Observatories.* 1970, XI + 120 pp.

22. L. N. Mavridis (ed.), *Structure and Evolution of the Galaxy. Proceedings of the NATO Advanced Study Institute, held in Athens, September 8–19, 1969.* 1971, VII + 312 pp.
23. A. Muller (ed.), *The Magellanic Clouds. A European Southern Observatory Presentation: Principal Prospects, Current Observational and Theoretical Approaches, and Prospects for Future Research. Based on the Symposium on the Magellanic Clouds, held in Santiago de Chile, March 1969, on the Occasion of the Dedication of the European Southern Observatory.* 1971, XII + 189 pp.
24. B. M. McCormac (ed.), *The Radiating Atmosphere. Proceedings of a Symposium Organized by the Summer Advanced Study Institute, held at Queen's University, Kingston, Ontario, August 3–14, 1970.* 1971, XI + 455 pp.
25. G. Fiocco (ed.), *Mesospheric Models and Related Experiments. Proceedings of the 4th ESRIN-ESLAB Symposium, held at Frascati, Italy, July 6–10, 1970.* 1971, VIII + 298 pp.
26. I. Atanasijević, *Selected Exercises in Galactic Astronomy.* 1971, XII + 144 pp.
27. C. J. Macris (ed.), *Physics of the Solar Corona. Proceedings of NATO Advanced Study Institute on Physics of the Solar Corona, held at Cavouri-Vouliagmeni, Athens, Greece, 6–17 September 1970.* 1971, XII + 345 pp.
28. F. Delobeau, *The Environment of the Earth.* 1971, IX + 113 pp.
29. E. R. Dyer (general ed.), *Solar-Terrestrial Physics/1970. Proceedings of the International Symposium on Solar-Terrestrial Physics, held in Leningrad, U.S.S.R., 12–19 May 1970.* 1972, VIII + 938 pp.
30. V. Manno and J. Ring (eds.), *Infrared Detection Techniques for Space Research, Proceedings of the Fifth ESLAB-ESRIN Symposium, held in Noordwijk, The Netherlands, June 8–11, 1971.* 1972, XII + 344 pp.
31. M. Lecar (ed.), *Gravitational N-Body Problem, Proceedings of IAU Colloquium No. 10, held in Cambridge, England, August 12–15, 1970.* 1972, XI + 441 pp.
32. B. M. McCormac (ed.), *Earth's Magnetospheric Processes. Proceedings of a Symposium Organized by the Summer Advanced Study Institute and Ninth ESRO Summer School, held in Cortina, Italy, August 30–September 10, 1971.* 1972, VIII + 417 pp.
33. Antonín Rükl, *Maps of Lunar Hemispheres.* 1972, V + 24 pp.
34. V. Kourganoff, *Introduction to the Physics of Stellar Interiors.* 1973, XI + 115 pp.

SOLE DISTRIBUTORS FOR U.S.A. AND CANADA:

Vols. 2–6, and 8: Gordon & Breach Inc., 150 Fifth Ave., New York, N.Y. 10011
Vols. 7 and 9–28: Springer Verlag New York, Inc., 175 Fifth Ave., New York, N.Y. 10011

OHIO UNIVERSITY LIBRARY

Please return this book as soon as you have finished with it. In order to avoid a fine it must be returned by the latest date stamped below

JUN 2 8 1983